Paul G. & Edith Lyman-Shafiroff

'Gus' of his Royal Majesty's Merchant Marine

Ture Alfvegren

Boatswain

LILYSOHL
Massachusetts

'Gus' of his Royal Majesty's Merchant Marine

Printed in the United States of America

International Standard Book Number: 978-0-9794288-0-7

Library of Congress Control Number: 2007902071

Published by LILYSOHL
881 Stebbins Road
Otis, Massachusetts 01253-0497

Printed by A & A Printing
Tampa, Florida
Cover photo, 'Gus', and all text photos property of the authors, except where noted

This book is lovingly dedicated to our mothers

Lillian Sohl and Edith May Shafiroff

&

the sailors of our

Merchant Marine Services

"They that go down to the sea in ships. . . . .
see the works of the Lord and his wonders in the deep."

Psalm 23
Book of Common Prayer

'Gus' of his Royal Majesty's Merchant Marine

by

Paul G. & Edith Lyman Shafiroff

If

Rudyard Kipling

1865—1936

If you can keep your head when all about you
Are losing theirs and blaming it on you;
If you can trust yourself when all men doubt you,
But make allowance for their doubting too:
If you can wait and not be tired by waiting,
Or, being lied about, don't deal in lies,
Or being hated don't give way to hating,
And yet don't look too good, nor talk too wise;

If you can dream—and not make dreams your master;
If you can think—and not make thoughts your aim,
If you can meet with Triumph and Disaster
And treat those two impostors just the same:
If you can bear to hear the truth you've spoken
Twisted by knaves to make a trap for fools,
Or watch the things you gave your life to, broken,
And stoop and build 'em up with worn-out tools;

If you can make one heap of all your winnings
And risk it on one turn of pitch-and-toss,
And lose, and start again at your beginnings,
And never breathe a word about your loss:
If you can force your heart and nerve and sinew
To serve your turn long after they are gone,
And so hold on when there is nothing in you
Except the Will which says to them: "Hold on!"

If you can talk with crowds and keep your virtue,
Or walk with Kings—nor lose the common touch,
If neither foes nor loving friends can hurt you,
If all men count with you, but none too much:
If you can fill the unforgiving minute
With sixty seconds' worth of distance run,
Yours is the Earth and everything that's in it,
And—which is more—you'll be a Man, my son!

'Secret of the Sea'

'What pleasant visions do haunt me
as I gaze upon the sea
and the old romantic legends
in my dreams come back to me

Sails of silk and ropes of sendal
such as weaved in ancient lore
and the singing of the sailors
and the answer from the shore

Like the long waves on the sea beach
where the sand as silver does shine
with a soft monotonous cadence
throws its own lyric lines

And how I heard the ancient helmsman
chant a song so wild and clear
the sailing birds a slowly
poised upon the mast to hear

'till his soul was full of longing
and he cries with impulse strong
Oh Helmsman! for the love of heaven
would you teach me that wondrous song

Would'est thou, the helmsman he answered
learn the secret of the sea
only those who brave its dangers
comprehend its mystery

In each sail that skims the horizon
in each land with a blowing breeze
I behold that stately galley
and hear those melodies

Still, my soul is full of longing
for the secret of the sea
and the heart of the great ocean
and its ancient melodies'

Henry Wadsworth Longfellow

CONTENTS

Illustrations

All photographs are part of the authors collection of pictures Gus acquired during his many years at sea. Although not an avid photographer, Gus obviously had the desire to preserve these images for posterity. The following pictures and illustrations are credited to the following:

Page - 40-collection of Marvin Cech

Page - 60, 61, 62 ,80 & 129-134 - The Mariners Handbook-Inter. Textbook *Co.* 1911

Page - 139-Postcard *'S.S. Demosthenes'*, Geo. Thompson & Co. London

Page - 144-2nd Gordons & 2nd Scots Guards on board the *TSS 'Lake Michigan'*
www. Kinnethmont.co.uk./1914-1918_files/John Innes-Thiepval/troopship-oct-1914.jpg

Page - 146-British troops on Sulva Bay- "The Nation at War", W. John Abbot, New York 1917

Page - 151-Official U.S. Navy

Page - 250-Chart No.1,Nautical Chart Symbols, Abbreviations and Terms published by the U.S. Government Printing Office.

Page - 251-U.S. Coast & Geodetic Survey published in Washington, D.C. Dec. 1926

About the Authors

After telling my wife about Gus and the years I knew him as a boy, she encouraged me to write this story and through her persistence the work we did together over several years resulted in this book.

We searched through old trunks, dusted off photo albums, retrieved Gus's sailing logs and brought back to life again the adventures, fortunes and misfortunes of this man's life.

Paul received his Doctorate from the University of Massachusetts in 1974. For twenty–seven years he served as a law enforcement officer and as a Supervising Principal for the Southern Berkshire Regional District in the Berkshires, Massachusetts.

In 1966, he was commissioned a Lieutenant in the United States Army and served as an Infantry Platoon Leader and discharged as a Captain. He retired in 1995 and devotes his time to practicing the piano works of Frederick Chopin.

Edith took on the task of organizing all the material that went into the reconstruction of this fascinating story of a man and his love for the sea. Her combined interest in history and music was also the encouragement that enabled us to revive the music that was so much a part of a sailor's life: the Sea Chanty. Edith spends her time gardening and is an avid horsewoman.

Acknowledgements

This story would not have been told without the encouragement and support of first, my wife. Edith felt that this story needed to be shared as it reflects those precious values of integrity, forthrightness and loyalty that seem neglected in today's culture. Her sister Eva provided invaluable assistance in support of this book and as researcher. Anne Reiss assisted in it's initial editing. Thanks also to the technical support of Douglas Lyman and Daniel Lyman for providing the assistance in graphic design and technical manuscript preparation . . Thomas A. Consolati for his encouragement and support as former Superintendent of the Gloucester's Public Schools, Massachusetts. A grateful word of acknowledgement also goes to my brother David for his many hours of assistance in researching the vessels Gus sailed on and to Marvin Cech for permission to use his photograph. A special thanks as well to Nancy Hahn. To Jodi Pow, our love and gratitude for her work in the final editing and proof reading.

1

Up the Road

I remember vividly that day in the early spring of 1951. I was just nine years old and we set off to look at the 'farmhouse,' a country home my mom was interested in buying. It was hoped that our family would have a place to go to for the summer. Brooklyn was hot and a polio epidemic was raging throughout the city. Dad was a physician and saw new cases of polio spring up daily. This country home would reduce our family's exposure, and in Dad's mind, 'keep us safe.'

Our car bounced along and I held on tightly to the top of the seat in ready expectation of each bump. Up ahead, we could just barely see a red house, as we got closer, I could see big old barns. The country air was alive with sounds and smells.

The car suddenly dropped into a muddy hole and eased over like a listing ship. The tires spun as my aunt tried to get the car moving forward to no avail. Back and forth, the car rocked as she tried to drive it out. With each rocking motion, the wheels sank deeper; we were now really stuck in what seemed like the middle of nowhere.

We filed out of the car to see what could be done and immediately noticed that the rear wheel was buried in the mud hole over the hubcap. We all stood in place, eyeing one another, wondering if there was even a chance that someone would come along. We were 'city people' out in the middle of nowhere, standing around, waiting for some inspiration, hoping it would fall out of the sky.

It was a beautiful day with plenty of interesting things to look at. This took my mind off our dilemma. Stone walls lined the road; big fields edged with large old maple trees opened up toward the red house which was brightened into visibility by the glare of the morning sun. Beyond the red house and further up the road we could barely make out a large white farmhouse with wisps of white smoke circling above its little chimney. There must be someone living there and that gave us hope. Of course, we could always just wait, hoping someone would come along and pull us out or, I thought, we could walk up the road to the white house and ask for help.

After some thought, Mom selected the latter idea as being our best remedy. Having so much fun just taking in all the sights, I went along with the idea. After all what did I know at nine years old? I was so preoccupied with all that I saw looking up at the sky from horizon to horizon, a world so distinctly different from Brooklyn. There were flocks of birds fluttering among the trees and a stone wall lined the road. Where did all these stones come from? The huge open-meadow landscape was more than I had ever really seen and experienced all at once. Just as reality set in and I began to become more concerned with our dilemma, I heard a faint, jingling sound, similar to sounds I've heard in Macy's department store during Christmas. It sounded like it was coming from up the road, a shuffling and jingling in rhythm and a voice calling cadences.

Now I could hear the rattling chains and a clippety clop, clippety clop, then a deep voice rang out, "Come on, Dick, come on, Boy!" and echoed down the road as the noise came closer. Clippety clop, clippety clop as the horses hooves struck the stones in the road. Sounds of chains and squeaky leather accompanied the rattling and the entire band became more present by the moment. Then, horses, huge horses. Around their necks were large collars with shiny brass finials on the top and over their flanks hung draping chains.

I had seen tow trucks pull cars but I never ever envisioned a horse rescuing a car; it must weigh thousands of pounds. Well, maybe the horse could do it. These were very large horses. The closer they came the more I felt this lump growing in my throat and I didn't know if it was fear or just their awesome sight in such close proximity to my smallness at the time. I remember being at the Brooklyn Zoo, seeing elephants, but they were always in a cage and out of reach. Now I had to strain my neck to look up at these horses within arms' distance. A tall, wiry, old-looking man with white hair, pulling against leather lines hollered out "Whoa Dick, whoa Boy!" The horses came to a halt and stood looking at us through their blinkers. Their harnesses jingled as they shuffled in place. That smell, that sweet smell of horse hit me and I recalled my only other experience around a horse. At Coney Island, I had petted a police horse's nose. How soft it was coupled with the rich smell of leather tempered by the damp summer evening air.

Every once and a while, the horses would snort with a loud rush of air from their flaring nostrils, followed by a cloud of steamy vapor. The man driving the horses looked old and had a wrinkled sun-burnt face. He was dressed in old dungarees, a flannel shirt, and a pair of high-backed 'Keds' with the laces tied up around his pant legs.

"Stuck, are ya?" he croaked. My mom and aunt responded at the same time "Yes," looking very embarrassed.

"Gus is my name and this here is Dick and here is Boy", he said introducing the horses. "Let's see what we can do." He backed the horses up close to the car bumper, their hooves shuffling and shifting in the muddy road. Gus unhooked a chain hanging on the whipple-tree, leaned over the bumper to hook the chain around it. I thought for a moment, "does he know where to hook it?" I was compelled to interrupt what Gus was doing to tell him and I started, "Excuse me, but you can't hook the chain there. Let me show you." As I reached for the chain in his hand explaining as I went, "It has to go here on the frame." As if startled, whatever gentleness his voice had held before now disappeared and in its place a gruff, resounding bellow-like response was coupled with an annoyed look that made me shudder inside. "Hey mate! Did ya ever work a team of horses?"

I looked at him sheepishly and responded "no", as I shrugged my shoulders and glanced over at my Mom. I felt my face flushed red.

Then he followed with the bigger question "Hey mate! How old are you anyway?" That did it. I thought a moment, then I looked him right in the eye and said "I'm nine." I can still remember my attitude at the time and that I wanted to punctuate my answer with a little 'what about it?' Well, the way he was looking at me, his hands now boldly on his hips, crouching down a little! That stare from his squinting blue eyes pierced right into me. It was enough to make my knees knock. Gus then hooked the chain to the car. I know not where as I was too busy looking the other way. "Come on, Dick! Come on, Boy, heave to!" he bellowed to the horses as the slack in the chain tightened and then out came the car from the muddy hole like the *H.M.S. 'Titanic'* was being raised.

Gus unhooked his chain and suddenly hollered out again, "Come on, Dick, come on, Boy. Let's get a move on!" Off went the horses, the dragging chain and Gus. My mom didn't even have a chance to thank him but I waved and yelled out "Good bye" for all of us. He just kept on heading back up the road toward the white house. He did stop briefly, gave a hitch to his dungarees and in a strange gait sauntered up the road, never looking back.

"Who is that man and where did he come from?" I thought out loud. My aunt piped-up, "Who did you think he was, the Lone Ranger!"

We were on our way again. Just a little further up the road we drove into the yard of the red house. My mom and aunt walked through and around the house while I checked the barns. They were huge with old pieces of equipment strewn all throughout the barnyard. Mom seemed please with the house. "It will do" she said to my aunt. I couldn't help but look up the road as the sounds of the harness and chain now had faded away. Back to the car we went and the long drive back to Brooklyn.

Before I knew it, school was out and summer vacation began. I had anticipated this day for months and now we all, my brothers and sisters, Mom and Dad piled into the car loaded with bags of clothes and food and headed on our way up to the Berkshires. It was a long trip. Before we knew it, we were on that old bumpy road approaching our new country summer home. Our first overnight was a surprise nightmare. Between the heat and a combination of mosquitoes and moths, we were uncomfortably annoyed all night. The challenge was to keep our head under the sheets just long enough to dodge the bombers and time it right to surface for air before another onslaught.

The morning finally came, the sun shone brightly through the east windows. Our clothes were damp and wrinkled from the humidity. On this, our first day, my brothers, sisters and I were ready to go exploring. There were barns and meadows and all kinds of old curious equipment lying in the barnyard awaiting our inspection. I thought of how I could get away to visit the old man, Gus, up the road. I wanted to say hello and tell him I was here for the summer but Dad had a list of things to do and I figured I better take care of those first.

Toward evening, after I had worked myself to the bone, cleaning and unpacking, my brother and I decided to build a fire pit from the stones lying around the yard. We planned to cook hot dogs over an open fire and went around picking up kindling for the fire which Mom would light for us. The sun was getting low in the sky; a balmy summer breeze swept over the fields and our campsite. Mom started the fire and we all took turns throwing little sticks into it until we got a good bed of red-hot coals.

We had chosen a spot for the fire pit a little ways from the house. We had to walk through the edge of the field and soon we had a path stomped down. We sat by the fire watching the flames flicker when I suddenly felt this strange presence. I looked up, turned around and lost my breath for a moment as there standing right behind me was the man from up the road.

Gus was poised right above me, squinting and looking down at us, his hands on his hips with a very disgusted look on his face. His sleeves were rolled up and I could make out a tattoo of an anchor on his forearm. I couldn't imagine why he looked so mad.

If I wasn't mistaken, I did say good bye to him the last time I saw him when he pulled our car from the mud hole a couple of months earlier. I thought "mightn't he be happy to see me?"

In my mind he was so happy to know I came back, that he walked all the way down the road to visit us.

"Hi, Gus." I said, surprised he looked so contrary.

My brother, David, sitting next to me murmured, "Where did this guy come from?" We were all spellbound. The silence was finally broken after what seemed like an eternity.

You could hear a grasshopper hop, then Gus bellowed "You guys can't be walkin' on this grass; this is me hay!"

He made his blue eyes squint and dart back and forth, making sure we got the message. There was no doubt that he meant business. "You boys got that?" he said in a growl. With that he spun around in that strange gait, and with a jerk of his trousers he disappeared up the road. My brother, Mark, inquisitively chimed in, "Why does that guy walk so funny?" Nervously, I replied, "Why don't you run up and ask him?" Looking at him parade up the road, you would have thought he was marching, the way his arms swung back and forth, his body stiff as a board. It reminded me of when my mother became angry, except she added a huffing noise to the scene.

Mom and Dad opened the screen door and asked, "What's going on out here?" I explained to them about Gus not wanting us to walk on the grass because it was actually his hay. Confused, they shrugged their shoulders, looked at one another and went back into the house. Again, I focused my eyes up the road and could just make out Gus going into his house followed by a loud slam-bam which reverberated and echoed across the hill. We all remained by the fire watching the embers die down and Gus became the main topic of discussion for the remainder of the evening. My brother Henry thought Gus was a pirate because he saw tattoos on his arms. David thought he resembled 'Popeye', especially the way he walked, his funny accent and mannerisms. Mom surprised us with marshmallows and we spent the rest of the evening around the fire, roasting, eating and talking about our new country experiences. There is something very soothing about an open fire, sitting in its warmth, watching the sparks drift upward and just staring at the glimmering coals. It always seems to put me into a very thoughtful state and that night my mind went drifting back to thoughts of my grandfather who had passed away almost a year before. Grandpa Oscar was a rugged character, a captain on a merchant ship. When I visited my grandparents on Long Island I would always seem to get in trouble with him. He kept to himself, never seeming very friendly. He was distant and commanding, so unlike my grandmother who was very dear to me. To make up for his lack of affection she would show me his 'sea treasures' his naval sword and hat worn in the war. She seemed to want to reassure me of his love and to explain his distance simultaneously.

On a visit my grandparents made to our house in Brooklyn, Grandpa unexpectedly sat me down at the kitchen table and described to me with a drawing how a capstan worked on a ship. I felt so honored. Unfortunately, several months later, he passed away—I wished I had gotten to know him.

Lt. Oscar Pihlack at the helm of his destroyer in WW1

What is it, I thought, about men of the sea and what is my interest in seafaring? There was so much more I wanted from my grandfather. So much I felt drawn to. Yet the information was difficult to obtain from him and then he died. I was seeing in Gus the very same characteristics.

They were overbearing, kind of condescending and haughty all at once. Yet, I just knew there was some inner wisdom they both possessed and seemed to guard very closely. I wanted to know. Who are these men? Maybe their knowledge was a secret, 'the secret of the sea.' Anyway, as the summer progressed, I decided I really needed to go up and pay a visit to Gus. Mom didn't think it was a good idea, maybe I would be a bother to him but I told her I knew he would want to say hello to me. I went up the road and found him outside working on his house. I said "Hi Gus. It's me!" He just kept on working as if he didn't hear me. I repeated myself, this time a little louder. He just kept on working along as if I wasn't even there. He looked like he was resetting nails in the clapboards and putting caulk over them. "Do you need any help?" I asked further. That got his attention. "No," he piped up in a loud growl. So, I just decided the best thing to do was to hang around and not bother him. After a while I started to get this feeling that he was getting annoyed like Dad does when you stare at him while he is struggling with something he's fixing. A few minutes went by and then abruptly and in a holler, "Hey Mate! Don't you have anything to do?"

"No, not really," I gave a good simple answer to a straight and simple question. I figured if he couldn't use my help perhaps he wouldn't mind answering a few questions. That was my logic at the time anyway. So, I posed my first question, "Are you a sailor or something?"

Clearing his throat like he had something stuck in it, he bellowed, "What do ya mean by that! Mate?" I restated my question thinking maybe he did not understand my Brooklyn accent. Gus answered back in a very gruff and accented voice, "Ya, what's it to ya!" I then said, "No, I mean when you were young?"

Gus now stiffened up, dropping his arms rigidly to his sides as if now clearly confused, "Well, I'll be. What ya mean, mate?" I put it to him again as clear as I could, "You know, when you were a kid?" "Cheese and rice," I heard him mumble, whatever that meant as he was now fumbling with his can of nails, dropping his hammer and scrambling to pick up his tools at the same time, yelling "Skipper, Skipper get on out here!" At that moment, a lady appeared carrying a pitcher of lemonade.

"Care for some lemonade?" she asked pouring a glass for Gus which he guzzled down as if going somewhere in a hurry.

"Thank you, I would." I said, as my throat seemed hoarse from trying so many ways to put my question to Gus so he would understand. This lady that Gus called Skipper also had an accent. She invited me in the house while Gus disappeared. We sat drinking until we finished the lemonade, talking about their farm. I began to ask her "Why does Gus call you Skipper?"

Well, she answered "It's a long story but I used to have a sailboat that Gus and I sailed; she was called the *'Nebula'.*"

"Do you think Gus would tell me sea stories?" I asked, a bit impatient by this time.

"I think he will but you gotta' give him some time to get to know you."

This was the beginning of my friendship with Skipper. We became good pals and when Gus would come in from doing barn chores to take a coffee break, I spent time with him and he became friendlier towards me.

Gus had his own favorite chair at the head of the table and if anyone, Skipper advised me, was found sitting in it, they would be done for. I respected that bit of information and claimed the chair alongside his as mine. My first formal introduction to Gus was made at the dinner table one evening. His first rule, however, that is before any conversation, was to challenge me to an arm wrestle. He would thrust his elbow on the table; spread open his big hand and say "Come on, mate, put 'em up!" Oh boy, I knew this one from Brooklyn.

There stood his arm atop the table with hand open, awaiting mine like a shark waiting for its prey.

I could clearly see part of Gus's middle finger was missing. It was like a stump sticking straight out aimed at me. I put my elbow on the table, leaned forward and grasped his hand as best I could. My hand immediately disappeared; it was lost in his huge hand. After a little pressure, he took a little from me in return and it looked like his hand may have quivered when I looked up at him. I saw a grin on his face like a smirk, his squinting blue eyes twinkling in the light of the kerosene lamp when suddenly…bang! I heard the bang at the same time I felt my knuckles smack against the table top with a thump.

Of course, we played this game as kids in Brooklyn. It seemed everyone would arm wrestle and try to see who was the strongest. There were tricks to this game like how one positioned one's arm on the table and how one grabbed hold of the opponent's hand. The smarter kids would also press their thumbs against the back of the opponent's hand and put pressure against a tendon, hoping to handicap the other to defeat.

In Gus's case there was no way to apply any of these techniques as his hand was too big and his strength overwhelming. Although, I must admit I did try a few of these tricks with him. It gave him great pleasure to play around with me. Sometimes he would make believe he was going down and when I thought I had him, he would quick as a flash, flip my arm over. Then a big, deep resonant laugh would echo off the kitchen wall, his mood would change and he would become more talkative. Now he had established with me that he was the bos'n. Thus our arm wrestle became a routine, a greeting before any conversation began. I spent many evenings with Gus and Skipper around their table. Little by little we became friends, yet still it took awhile for Gus to loosen up and share with me his sea adventures despite my eager persistence. Every evening began the same way; this was our ritual—first the arm wrestle and then the questions which seemed to be his approach to a story. He invariably would ask me "Do ya know….?" this or that or "What do they teach ya in school?"

From time to time I would get a taste of his temper which was quick and deliberate. It quickly became anchored as a lesson for me. His displeasure was something I really wanted to avoid, not that I didn't need a lesson or two.

On one occasion I needed a shovel to help Mom dig up her vegetable garden. I went up the road to ask Gus if I could borrow his. He readily gave it to me and I marched back down the road with the shovel over my shoulder and spent the afternoon turning sod over. After dinner I finished up the garden but instead of bringing it back right then and there, I set it against the house intending to take it back in the morning. The next morning I noted there was heavy dew as I quickly carried the shovel up the road and handed it to Gus. He looked at me in that funny way of his, his eyes squinting. He then let out a series of expletives as he increasingly raised his voice. Evidently the shovel had started to rust from having been left out all night in the dampness. He was livid. I didn't know what to say. Gus proceeded to make it quite clear that I was to always return anything I borrowed in the same condition it was loaned. As far as Gus was concerned, I was totally irresponsible and he had no problem letting me know that: another anchored lesson.

As for expletives, Gus would cuss whenever he began to lose his temper. He never used vulgar words like people use regularly today. He did, however, in honor of never taking the Lord's name in vain say, "Cheese and rice"

I witnessed another example of his temper when we were out cutting firewood one morning in a brushy meadow beside the house. He had his double bit axe and was chopping away while I was stacking the trimmings when a shot rang out loud and close. It was hunting season and Gus had no particular penchant for hunting. He stopped chopping and yelled out "You bloody fool, get on out of here!" As he continued yelling he tromped through the brush in the direction from which the shot had been heard; his utterances reverberating in my ears. When he spotted the hunter, he raised his axe and started running toward him. The hunter quickly took off with Gus gaining on him. After a short way, Gus let out one last bellow, "Don't come back, you bloody fool or I'll tan your hide."

Gus was not a person to mess with—he would jump right into any situation before you could blink an eye. He didn't stand for carelessness or irresponsibility. Not that he would have hurt that hunter. However, he felt it to be his duty to let him know how irresponsible he was. Although it may appear rash, he felt it necessary to deal with the intrusion right there and then in the only way he knew how to be effective.

My mom told me of a similar situation with my dad when he was driving to his office down Flatbush Avenue just before the approach to the Brooklyn Bridge. A garbage truck cut him off and when they got to the traffic light and had to stop, Dad got out of the car, stepped up on the running board of the truck and started yelling at the driver through his open window. With that, the driver hauled off and punched Dad right in the eye causing him to fall off the running board. The light changed green and Dad staggered back to his car and proceeded on his way to the hospital to see his patients. Mom had to put a steak on his eye when he got home that night..

It all seemed customary behavior in those days. If someone offended you, it was dealt with quickly and directly. Today things are different. One tries not to consider such reactive behavior because the repercussions could be more serious.

Neither Gus nor my dad had violent tendencies; they believed they had to stand-up for themselves when wronged. Imagine being a boatswain (bos'n) on a ship and having to earn the respect to gain authority over a crew by example. Your job is to pass down orders from the captain and/or first officer and then be responsible for seeing to it that all orders are carried out. This job takes a stern, confident person with strong leadership skills and mastery of the skills to keep a crew in line. I'm sure there was little room for compromise or exception; every situation would have to be dealt with in a firm and consistent manner.

2

By the Kerosene Light

Gus had several possessions that he truly valued and prized. One was a copy of the 'Old World Atlas' kept in a sea trunk in his little room off the kitchen. Here he kept all of his very guarded treasures. I would become familiar with these treasures and their stories as time went forward. After a, "Do you know…?" question, Gus would push back his chair and in a determined manner retrieve his atlas and proudly place it on the table. The gentle, soft glimmer of the kerosene light would reflect off the atlas' glassine pages as he searched for the place he wanted. His index finger would point to some obscure location while his other stubby finger always obstructed the name. His first story took me to Gotland, an island off the coast of Sweden. A place I had never heard of. Then pointing to a place he called Wisby, he explained, "This is where I was born, mate."

The kerosene light flickered as it caught a draft making it difficult at times to see the detail on the map. The gentleness of his voice and the rhythm of his accentuated English would lure me into a sense of feeling like I was back in time, actually transported there amongst the people and places he described. In a proud voice, Gus told me that Gotland was where the Vikings came from. He described his village and the harbor where he was born, its quaint and medieval buildings and marketplaces.

Gus told me as a child he spent much of his spare time on the docks watching the sailing ships as they entered and departed Wisby's harbor. He said he was always mesmerized by the white sail sheets, the masts and the clamor of the sailors as they ran up and down the tall riggings. He found school very boring and took every opportunity to go down dockside to watch and listen to the sailors as they sat around spinning yards, singing sea chanties and working aloft.

Gus said his father was very strict, insisting he go to school. This did not stop Gus from his longing for adventure, nor did it quell his desire to go to sea. Nothing else mattered and his father's anger only pushed him further into the world he had created for himself. Despite his many scoldings and punishments, Gus looked to the sea as his hope "So anyway, finally, one morning," Gus went

Gus explained, " I just couldn't stay away. It was so exciting listening to the tales of the sailors sea adventures, singing chanties and watching all the different kinds of ships sail in and out. You can tell the type of ship by the sails they carry and the number of masts. Looking at a schooner you would see usually two masts. The after mast and the main each have a top mast and carry a gaff mainsail. The way the foremast is rigged determines the different types. Now a brig also can be a two- masted ship and was the most common one we'd see come in to port. The difference between a brig and a bark is that the brig carries a large mainsail. A brigantine has no square sail. Its main mast has two parts with a square topsail. Lookie here! This should explain it better. These are pictures of the different common types of sail ships," as Gus flipped through loose pages in the back of his Atlas.

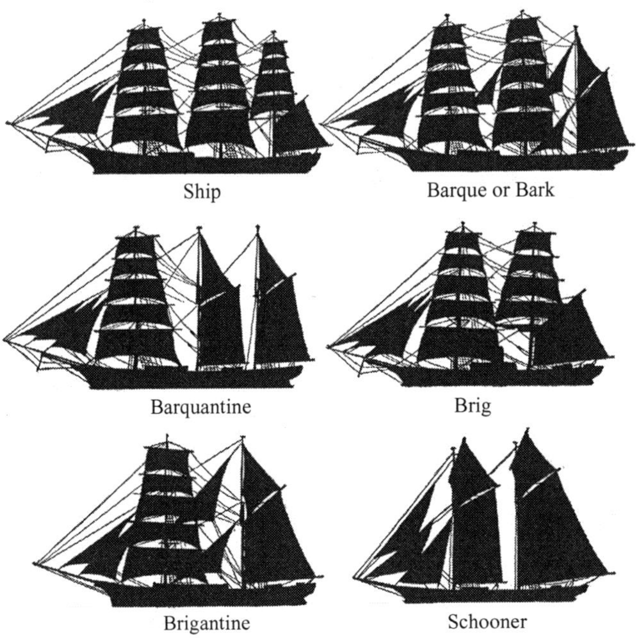

“These are things you should know also as it will help you understand the language I use when I talk about ships. There are two types of rigging. One is called standing rigging and that’s where the masts and the bowsprit are connected to each other by shrouds and stays. Shrouds and stays are words used to describe the type of rope because ropes aren’t just ropes. They all have purposeful names. Each rigging has a name like halyard, brace, shroud, stay and lift. Well, wherever a line connects to a mast and the bowsprit, we call that standing rigging. When a line is attached to a sail and yardarms or blocks, we call that running rigging and it is generally slack. The standing rigging is mostly taut.”

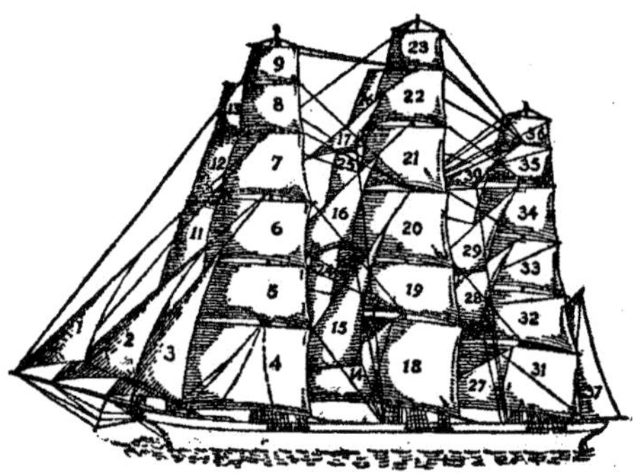

Full-rigged ship under all plain sail

1-flying jib; 2-jib; 3-fore-topmast staysail; 4-foresail; 5-lower foresail; 6-upper fore topsail; 7-fore topgallant sail; 8-foreroyal; 9-fore skysail; 10-lower studding sail; 11-fore topmast studding sail; 12-fore topgallant studding sail; 13-foreroyal studding sail; 14-main staysail; 15-main topmast staysail; 16-main topgallant staysail; 17-main royal staysail; 18-mainsail; 19-lower main topsail; 20-upper main topsail; 21-main topgallant sail; 22-main royal; 23-main skysail; 24– main topmast studding sail; 25-main topgallant studding sail; 26-main royal studding sail; 27-mizzen staysail; 28-mizzen topmast staysail; 29-mizzen topgallant staysail; 30-mizzen royal staysail; 31-mizzen sail; 32-lower mizzen topsail; 33-upper mizzen topsail; 34-mizzen topgallant sail; 35-mizzen royal; 36-mizzen skysail; 37-spanker.

“Also a rope is not tied like you would say. It is belayed. This way it can be released quickly. Most running rigging, whether it is through a block or runs directly down to the deck, is made fast to pin-rails located on the side of the deck rail. The rope is twisted around the pin-rail which keeps it fixed yet able to be released quickly if need be. A ship can have more than three masts as I'll show you later but she must be all square rigged. Also the masts are made up of three parts; a lower mast, a top mast and a topgallant. A barque (bark) may have four masts but the after mast must be fore and aft rigged and have no square sheets."

"A brig, usually two-masted is typically a smaller vessel. The barquantine has three masts with square yards on the fore mast, the main and mizzen mast with both fore and aft sail. A square rigged ship we called a windjammer or hooker."

"British vessels were usually called Limejuicers because the crews sucked on a lot of limes to prevent scurvy. Maybe that's why Englishmen are referred to as Lymeys."

"By the way originally everything used to connect anything to anything was made of rope but when the vessels got bigger, the rope for an anchor for instance took up so much space below deck because of its thickness they then went to chain. Put that in your pipe and smoke it."

"There are also other kinds of ships as well that I haven't mentioned. Some we call fore and afters, like cutters sloops and yawls. There are many derivations of the schooner types. The ketch and yawl have two masts. The cutter and sloop but one."

"Then there are the different classes of steamers ranging from passenger ships, cargo ships, tramp steamers, which carry cargo and passengers to petrol and coal tankers."

"So over time sail ships were replaced by these steamers but in some industries like fishing they are still used. You'll also find sail used in yachting and pleasure craft. There are still a whole bunch of sail ship buffs out there who keep the skill alive."

on with his story, “I took the few belongings that meant something to me and, instead of going to school, I went down to the docks. I had made up my mind that I was leaving home for good. I saw a captain standing on the gangway of a large frigate and nervously approached him. “Sir,” in as confident a voice as I could muster, “I’m looking to sign on for work aboard your ship.” There were always opportunities to sign on because many times sailors would go on shore leave and not come back, either getting drunk, running off or whatever, so I felt I had a good chance to get hired on. The captain kept looking down as he was checking off his cargo on the bill of laden, although I knew he was listening. His eyes occasionally would peer out from under his cap brim and then he finally asked me “How old are you, son?” to which I replied an untruth without hesitation, ‘Sixteen, sir.’”

“Have you been to sea before?” he inquired further as he continued checking his list, his eyes darting back and forth trying to do two things at once, which was making me nervous. “No,” I said, “but I’m willing to do the work.” The captain dropped his papers to his side and I could see he was thinking. After what seemed like a long time, he turned his head toward me, looked me square in the eyes, as if right through me, paused and then said “Get below to the galley and see what the cook needs! We’re getting ready to shove off, so be quick about it. There’s plenty of work to do here.”

“I headed up the gangway, across the deck toward the forecastle where I could see the galley pipe. I went down the ladder and found the cook standing by the stove. He was a rugged, short man who carried a knife tucked into his apron straps. There were piles of vegetables on a butcher’s block, pots steaming away on the galley stove with him running back and forth trying to attend to all he had to do. He stopped suddenly, looked at me and asked in a kind of gruff, annoyed voice, "What brings ya down here? Can't you see I'm busy?"

"The captain sent me down to work in the galley. He told me to see the cook." He looked at me, sizing me up to see if I was worthy enough to work in his galley. Finally, he said 'I'm the cook. What can you do, mate?,' I said to him, "I can do anything you need." 'Okay, mate,' he piped up. 'You can start with cleaning those kettles. When you're done, go down in the stores and bring up some flour.'"

"So began my first sea voyage. She was a fine ship. Although I spent most of my time below decks in the hold or galley I could hear the bos'n hollering out his orders, blowing his whistle and the crew making ready, casting off and setting out to sea. Occasionally, I would have to go on deck to bring the captain or first mate coffee when they called down for it. I could see sleek ships sailing by, all the same kinds of ships I saw as a kid when they entered the harbor in Wisby. Now I could see them in full sail, their bows cutting into the water, their wakes following behind. It was my greatest thrill to be at sea. I felt at home on the open water."

"As we sailed out to open water, I could see the waves break across the bow, sometimes sending water across the deck and even down the galley pipe. I could hear the sails as they flapped in the wind; we were now on the open sea. The crew was busy scampering back and forth across the deck while others were aloft on the spars. It made for a very busy time aboard ship."

"The galley was a hot place to work in. It was small and hard to move around the quarters. It was stuffy and I had to keep the fire going in the cook stove while running down to the hold for firewood. This kept me busy when I wasn't cutting up veggies and cleaning the pans. The cook did all the cooking and would throw out commands such as 'Give me this, get me that.' You know, I was a gofer. It was a job anyhow. I was at sea. My encounters with the captain and the crew were infrequent as the galley work took most all my time. The rolling of the ship made me sick and unsteady depending more or less on what the cook was making. Some of that stuff smelled much different than what my mother used to cook. Well, anyway, I got used to the 'heave and ho' and it was just a matter of time until I got my sea legs under me."

"My bunk was right next to the galley. I slept on a hammock which was stowed away during the day; everything onboard had its place and we all had to be sure it was there in the right place if someone needed it. Lord help ya, if it wasn't. It took a little getting used to—sleeping on a hammock. It was made from canvas and ya sort of was wrapped in it; you couldn't roll over without bending like a pretzel."

"I would have to get up at night to make sure the fire hadn't gone out and there had to be coffee for the night watches. Sailors were fussy about their vittles, 'chow' ya know. It had better be ready on time and plenty of hot coffee to boot. There was no room for mistakes or burnt food. Sometimes when the seas were heavy, I would run coffee up on deck. It was hard to carry mugs of coffee across water-spewed decks with the spray coming up across the bow; once I learned the 'tricks of the trade' so to speak, my job became easier on me."

"One night, a gale wind came up and the ship heaved and rolled violently. The crew was called to their stations. There was much running about on the decks and climbing in riggings securing everything down. I could hear the swish of the pumps as some of the crew were down below bailing out the bilge. The sea spray washed over and swirled around across the deck. Sometimes the spray would come down the galley pipe and drown out the fires. Crew members struggled across the deck, holding on to guy ropes, bodies bending against the wind and water. The helmsman was wrestling with the wheel to keep the ship heading into the weather as she heaved and rolled. I was trying to keep all our gear in the galley from flying around. It was quite a storm. You can't really imagine it unless you experience the fury of the winds and the high seas, sometimes twenty or thirty feet high. By morning, the winds had died down, the swells settled into rippling whitecaps. The sea gentled itself and we were able to resume our normal duties. Some of the crew was aloft repairing and adjusting the sails. Others were sitting on the forward hold using awls to stitch those sails that were torn. Rolls of rope were coiled on the deck to dry out in the sun. 'Coffee for the captain at the helm, right away,' the first mate bellowed down the galley: coffee, coffee, all day and all night long. I was hoping someone would take me on, show me how to work up in the rigging—but when? I didn't know when but I really wanted to work with the deck crew."

"One evening the cook ordered me to take the captain's supper down to his cabin. I went below toward the stern where his quarters were located, knocked, waited to be told I could enter and then set his supper down on his table. There were all kinds of charts and instruments, which the captain used to plot the ship's course. The captain was studying a chart when I came in. This was the first time since we set sail that I had the opportunity to see the captain alone. He looked up at me and asked, 'How do you like working down in the galley, mate?' I was surprised by the question and responded quickly with a line

I had long rehearsed, "I like it but I hope someday to be able to work up on deck with the crew." The captain nodded and with an enthusiastic tone to this voice he said, 'Well mate, we'll have to see about that.'"

"The following day, the boatswain peered down through the galley way and yelled 'Hey Gus, up on the deck, right now!' I sprang to my feet and up I went forward to where the boatswain was now standing, his hands on his hips, his feet well anchored on the deck and his chest puffed out. Standing next to him was an older crewman. He was leaning on a rope just gazing off into the horizon not seeming to pay any mind to what was going on. 'Gus', the boatswain blurted out 'this here is Johnny, the lamp trimmer. The skipper has assigned you to work with us now up here on deck. Johnny will show you what your duties are. Be sure you listen carefully and pay close attention to exactly what he tells ya, got that mate! I don't want any mess up on my deck, ya hear!'"

"Aye, Sir!" My heart felt like it was stuck in my throat. I'm not sure if it was the excitement of getting assigned to deck duty or the bold presence and bravado of the bos'n. This was what I wanted and felt I would do my very best to show the skipper how well I could handle this new job."

"Johnny had this bulge on the side of his cheek that moved back and forth from side to side. He gave me the once over. He didn't seem at all excited about having to take me on. He turned away, leaned over the rail and let go of whatever he had in his mouth. Later I would find out that's what sailors call a 'thumpin' quid,' a wad of chewing tobacco. He turned, looked at me in a kind of suspicious way, set back a little, gave a hitch to his trousers, and then he said to me, 'You do what I tell ya, don't get in me way and everything will be okay.' I trailed along behind him as he walked down the hatch below to the crew's quarters. It was one big cabin with some hammocks stowed up against the planks under huge wooden deck beams. Some of the crew were asleep all bundled up like in a cocoon, probably the night watch. The cabin was dark and damp like a dungeon. 'Here, mate, this is yours. Stow your gear here, make sure you watch over it,' Johnny emphatically recommended."

"Some of the crew popped their heads from under their bedroll and stared over at me. These were all veterans of the sea; it was like being a little frog in a big pond with all big frogs."

"I stowed away my gear, what little I had in my cubby and unraveled my hammock. I slipped into it, hoping to catch forty winks. I fell asleep but was only to be awakened by the bos'n blowing his whistle and then hollering down the hatch 'All hands on deck and be quick about it!' The crew members in their hammocks rolled out into their shoes laid out below and ran lickety-split up the ladder on the deck. I followed, but wound up last up the ladder and, of course, last to get into formation.

Johnny never told me about where to stand in the line. After a little scampering around everyone managed to find their place except me. When I finally found a spot, the bos'n walked the line and abruptly stopped in front of me. With a thrust of his neck he was in my face. 'What took you so long, mate? Be quick about it means right away. There's no time for lolli-gagging around down below. You got it, mate?' My face was now flush red as the bos'ns nose pressed against my ear. He muttered, 'We'll make a man out of you yet.'"

"I could hear chuckling coming from the ranks when the bos'n backed away. With one last stare, he hollered out, 'Now to your stations! Step lively, lads!' I looked around and found Johnny, approached him and waited for him to tell me what to do."

"Then abruptly, he said, 'Follow me, mate' as he turned and went forward. As we passed the main mast he turned and with a heave, pushed me against the mast. With those piercing eyes of his, he glared at me and muttered, 'I don't take kindly to having you tag along with me and I don't need the evil eye from the bos'n. When he says, "on deck," that means pronto, if you catch me drift, lad.' I was sure off to a poor start. I had thought the crew would be happy to have another hand on deck. This wasn't turning out to be my day; whatever excitement and joy I felt was slowly dwindling. Nevertheless, I stayed on my toes, was attentive to everything going on and wasn't waiting for anyone, not even Johnny to tell me what I had to do. If I saw something that needed attending to, I did it. There wasn't going to be a next time. Later, towards evening, Johnny sat me down and in a gentle voice explained to me the facts of life onboard. 'Mate, there are certain things you have to be able to do as a crew member. You have to be always on your toes and respond to orders right away. This is dangerous work and one mistake could mean someone's life. The bos'n is the boss of the crew, the captain and first mate run the ship, plot the course and make the big decisions. The bos'n, as I said, is the crew boss and when things don't go right, he has to answer to the captain. The captain then comes down on the bos'n and then he comes down on us; it always runs downhill. So mate, be steady and keep your eyes and ears peeled.' With that, Johnny left."

"While thinking all this over, I gazed out over the bowsprit and watched the waves gently play against the dipping and rising bow. I thought of home. The water was calm. I hung my feet over the bowsprit and just took in the gentle breeze as we were making way. It was a beautiful day and I felt so much better since Johnny took me aside."

"The next day was also calm. Soft breezes whispered in the sails as they flapped against the masts; the sea spray tingled against my face. I felt good out here at sea. I thought about home and what my parents might be thinking after they found my note telling them that I went off to sea. I thought about my mother, how she must be worrying about me. I resolved to send her a letter as soon as the ship docked."

"One of the crew started to sing a song with a chanty rhythm, keeping pace with the lads who were hauling down a line from the mizzen. It was one of those chanties I heard when I used to sit by the dock. The crew would pipe in to sing the chorus while the chanty man sang out the verses. It added excitement to the work and just helped everyone feel good. You know, the work was hard."

"'Mate!' I heard bellow out from between choruses. I jerked my head toward the call as the bos'n pointed toward me. 'You! Up the rigging to the main spar and look lively!' Looking up, I grabbed on the shroud, stepped on the ratlins and up I went, hand over hand."

"Two crew working on the main spar handed me a line and told me to carry this line to the forecastle. I ran the line down the ratlin' and took it forward where another crew member was working another line. Together we made it fast and secure. Boy, did I feel good, finally doing something right aboard this ship."

"After a short time passed, those of us not on watch went back to our quarters, sat around chewing the fat, waiting for the dinner bell to ring."

"Here and there comments were made about what a good job we all did handling the mast. Words of praise were even directed at me about how well I handled that line down the shroud."

'We certainly fixed that straight-away,' I heard Johnny proudly say. This was a great crew and many lasting friendships were made here."

"Suddenly a cry from the crow's nest echoed 'Ship off the starboard!' We all went topside and stared over the starboard rail hoping to catch a glimpse. Life was lonely at sea and it was a welcome sight to see another ship every now and then."

"After a while I got to know the different merchant ensigns and was able to recognize their country of registry."

"Ships had many different sail configurations and I learned their names brigs, schooners, barks and even clipper ships."

" Boy, those ships were really something to behold under full sail."

“As the ship sailed by, we could make out four masts. She swiftly and elegantly glided through the water on our starboard. She was four-masted, rigged, long hulled and looking as though she would cross our bow. From the bridge, ‘Quarter to port!’ the captain ordered the first mate at the helm.”

‘Turn to port one quarter’ and with a flick of the wrist the first mate altered our heading. From her foremast we could now see her colors, the Stars and Stripes flapping in the wind. ‘Good speed,’ our captain proudly praised as we turned to port. Some of these Yankee blokes were real funny when they were sailing big, swift ships like that. You had to make way; they were too big to turn quick. When she passed we could see her sailors waving their hats and shouting God knows what when, ‘Ahoy, mates’ rang out from our crew; she was one of ours.”

“The captain grabbed the bull horn and yelled out across the water in the direction of the ship. ‘Where are you bound?’ ‘Hong Kong!’ echoed back.”

“Ahead of a good wind she raced on by. Her wake was huge. Our ship rocked back and forth swaying to and fro’ with the waves, then she silently disappeared as fast as she had appeared. We all went back down below and some of the crew started recollecting their own experiences onboard different vessels including stories about their having served in the Queen’s Navy. I was all ears; I now felt that I belonged.”

“When evening chores were done, Johnny called me forward and pointed up at the sky. ‘See those stars, mate; that’s the big dipper.’ The stars were brilliant that night; so bright they glistened in the water. Looking up, there it was— a handled dipper clear as a bell amidst the multitudes.”

“Sailors spoke often of the stars and had a certain fascination with them. From time to time, I saw the captain taking a fix on the stars with an instrument called a sextant. Johnny helped me a great deal in understanding all the different names sailors had for stars and constellations. He pointed to another star, ‘That’s the North Star. Someday I’ll show you how to reckon a ship’s position using the stars to determine your latitude and longitude.’”

“’By the way, mate , what’s your name? I know the crew calls you Gus but what about the rest of your name? What kind of name is it anyway?’ Johnny inquired further. “It’s Swedish. My real name is Ture Alfvegren and I was born in Wisby, ya know, in Gotland.”

“Johnny recognized the place. ‘I was there a couple of times; that’s where the Vikings come from, they say. Do you have any family?’ Johnny asked.”

“Yes, my parents and sister, I responded. ‘Well, that’s good,’ Johnny’s eyes a little kinder now. ‘So you want to be a sailor, that’s good’ Johnny went on. I looked him right in the eyes, "Yes, I do!" I wondered if Johnny could tell maybe that I wasn't sixteen. He never asked and another rule I learned was sailors don't meddle in anyone's business other than the job they have to do."

"I asked Johnny where we were sailing to. He told me he thought we were on our way to the Caribbean Islands, a place called Barbados. There was cargo we were to pick up and bring back home, sugar and molasses he thought."

"We would be arriving in about two weeks if the weather held out. I was getting tired so I headed down toward the crew's quarters in the forecastle. I drifted off to sleep, the sounds of the waves bumping against the hull. The creaking of the timbers and the sleepy sounds of the sailors were like heaven to my ears. We had many days of good weather and calm seas. I could tell we were entering warmer currents. Clouds were high and occasionally, sea gulls would circle overhead looking for tidbits that the cook threw overboard."

"Sometimes, we'd spot dolphins swimming alongside the ship dipping up and down in the bow's wave. One day we spotted a huge sea turtle just beneath the surface swimming along, his head broke the surface and he stretched his neck upward as if to get his bearing or just maybe to say 'Ahoy, mates,' then slid back beneath the surface and went his way."

"Sailing was an exciting experience and the sea was a life of its own. Everything was new, yet mysterious and I felt akin to this world. The jobs on deck were fun especially swabbing the deck from one end to the other, polishing all the brass fixtures and helping to keep everything onboard shipshape. After all, this was now my home, I took great pride in my work. I enjoyed repairing ropes, making cable and braiding them. There were other jobs too that I had yet to learn . 'All in good time,' Johnny would say."

"Sometimes I would go aloft stretching out on a spar just to feel the breeze in my face and watch the ocean ramble along. This was my greatest pleasure. I looked down toward the stern, watching the water as it swirled in our wake. I thought with pride, here I am now, a sailor, a wooden ship sailor."

"The next morning we spotted three small fishing boats tossing about in the waters as they tended their nets. I thought we must be near land since sea gulls were dotting the sky, swooping to get their share of the catch. The bos'n ordered the hatches forward and aft opened; the fresh air and sunlight rushed into the dark cavernous holds. The first mate took a party below decks to check out the cargo."

"Aloft from the crows nest came the call, 'Land ho!' A small craft was approaching, a sloop came along side. Who is that, Johnny? I asked.

"Oh, that's the pilot. He'll be steering us in through the channel to make sure we don't go aground," Johnny answered.

"These ports have narrow channels and if the captain is not careful, we can be in trouble. That's why these ports have pilots; they know these waters like the backs of their own hands,' he added further."

"Off the bow one of the crew was taking soundings with a line, calling the readings to the captain while the helmsman turned the wheel over to the pilot. The sounding line has knots every six feet and that distance between them is called a fathom.

The number of knots the crewman counts multiplied by six feet is the depth of the water. The captain knows the depth of his ship's displacement to keep from not running aground, ya know, not getting stuck."

"The bos'n then ordered all hands on deck to go forward and make ready to enter port. Our station went aloft to take our positions on the spars while others stood on deck readying the cables. 'Furl the mainsail!' the bos'n piped up. First the mainsail and then, the top sail then all square sails were furled, the only remaining sails unfurled were the jibs and spanker. This was a tricky job for the captain as he had to decrease the sail surface and speed as we approached the wharf. We had to be on our toes and ready for whatever changes were needed with the sail. We glided in like a pussycat on its toes. The deck crew cast off the cables to the shore-men who made the lines fast and snug. The crew aloft let out a big cheer as we made port."

"The port was very busy and congested with ships. The piers were full and there were many ships in the harbor anchored or moored."

"Entering the harbor was another schooner. I had never seen one that close up before. She was still under full sail and I could just barely make out the crew scampering about as they started to furl their topsails. Maybe it was the schooner we all saw at sea. She could be stopping in Barbados to take on stores before proceeding on her way around Cape Horn."

“In the wharf was every kind of ship imaginable: two-masted, three-masted, even a clipper ship along with fishing scows. Sailors and fisherman were running around, horses and wagons drawing away cargo, booms hoisting cargo and chants of bos’ns giving orders. This was my first opportunity to be part of the real action. I had always sat on the dock at home, watching the sailors work; now I was on ship watching the work on the docks.”

“Flags of many nations flew atop the masts. A crew alongside piped up a chanty as they unloaded their cargo; it was ‘Whisky for me Johnny’. I heard that chanty from our chaps but to hear another crew singing and working inspired our chanty-man to pipe up one of our own favorites, ‘Hanging Johnny’. We all got right into it, our chanty-man bellowing out the verse with us coming in on the chorus. We heaved our cargo all afternoon, setting it down on the dock. Blocks and tackles were swinging from the booms as we sang in unison, one chanty after another.”

"Our captain was busy greeting two men on the gangway. They then went through the cargo, checking it against his bill of lading."

"Later than evening, a crew of teamsters loaded the cargo onto wagons and by midnight the crates were all carried off."

"Early the next morning, our routine work finished, the bos'n piped out 'All hands on deck. Captain on the bridge. All hands forward!' We scurried up from below, gathering around the bridge."

"The captain addressed us. 'You men can go ashore; we won't be taking on cargo 'til tomorrow morning. I want you mindful of your behavior. I want you all back by midnight. Anyone breaking the law, returning drunk or not on time will forfeit one-quarter of his wages.'"

"That's all!"' the bos'n piped up, "Dismissed!"

Four Masted 'Clipper' at anchorage

"We all went below to get ready to go ashore. Some of the crew knew this port well, but Johnny said he had never been to Barbados. We tagged along with the rest of the crew."

"It felt good to go ashore — so different to be on solid ground. In town we saw sailors crowding the streets dressed in various uniforms."

"In the marketplace vendors sold everything from soup to nuts. This is where we could get a chance to pick up any supplies we needed for our journey home, some tobacco, mail that letter to Mom, whatever."

"We visited an old fortress. There were large cannons mounted on the parapets overlooking the harbor. I could just imagine having to sail past this fortress with these guns firing at us. Johnny piped up 'well mate that was the way it was when the Spanish governed this island. Nothing could get by them.'"

"Although we Brits did manage to show them a thing or two, that was a long time ago. Those cannons haven't fired in centuries so, don't worry," "Johnny was filling in with a little history."

Three- Masted Schooner at her anchorage

Suddenly Gus jarred me from my 'sailor of the past' musings. I had been transfixed by his story, all sense of time altered in the flickering light of his lamp. Gus leaned across the table and turned down the wick of the lamp, the soft glow catching faint shadows on the walls. "Got to be up early in the morning; weather looks good and I gotta get the hay in. If you want to come by, I'll show you how it's done." Gus walked me to the door; I turned and appreciatively said, "Thanks, Gus, for that great story, I'll see you in the morning."

I should have remembered to check with my mom before I said okay to anything. Hopefully, she would let me help with the hay. I talked to myself all the way down the road, shuffling along in the dark. I could barely make out the shadows of who knows what, the swish I heard go by my head, the rustling I heard in the bushes. I marched along making as much noise as I could. I even whistled to make sure nothing was hanging around. The fast march turned into a jog. There ahead I could see the lights of my house.

Mom greeted me at the door. "I can't imagine what you could be doing up there for so long," "Oh Mom, it's really exciting, the stories Gus told me about when he first went to sea, how he ran away from home when he was thirteen and lied about his age." Mom had this expression on her face that just let you know what she was thinking. Her eyebrows would rise way up, her eyes turned to one side and with a slight huffing sound she'd turn abruptly and stomp away. I spoke up, "Mom, I need to help Gus tomorrow get the hay in, can I go?" Her response was "We'll see in the morning." "Oh great", I thought, I'll have to wait all night long to find out if I can go.

Leaning back against my pillow, I tried to imagine what it must have been like to sleep on a hammock, to feel the roll of the ship to hear the ocean beating against the hull. I had this mental picture: a ship with sheets of sails, the flapping of the sails in the wind. It gradually lulled me into a deep sleep.

I was up bright and early. The sun beamed through the trees in the grove, their shadows flickered across my window. Nobody was up yet, not even Mom. So I wrote her a note. "Dear Mom: Went to help Gus with the hay. Love you."

I ran all the way up the road not wanting to miss a thing. The birds were flitting in the trees and high grasses. These early morning sounds and smells that are so characteristic of New England were especially noticeable to me that morning. I knocked on the screen door, went in and found Gus sitting in his chair tying his laces. I never saw laces that long. He wore those high-backed sneakers with the 'Keds' monogram on the side; they looked funny because I never saw a size that large. Skipper came in from the woodshed and I greeted her "Good morning." She proceeded to put a couple of sticks in the fire, coughing away as a down draft filled the room with smoke.

"Let's go, mate." Out the door Gus went, with me tagging along and on down to the barn we marched. As we walked up to the barn, there was Porky, Gus's pig, waiting for his breakfast. "Porky!" Gus bellowed out, "How did you get out of your pen? You're hungry, are ya? Stand by, mate."

Gus slid open the barn door and we were met with a chorus of neighs, moos, clucks and baa's. It was breakfast time at the old barn. Gus started with the horses and their grain, then chickens, duck, lambs and finally, Bessie, the milk cow had her turn. He then pulled a little three legged stool off a peg, picked up a metal bucket, propped the stool alongside Bessie, greased his hands with some Crisco he kept on a shelf and started for milking. Swish, swish, the milk hit the empty pail. Bessie didn't seem to mind as she was too busy munching on her hay in front of the stanchion. I stood back as I read somewhere or heard on television that sometimes a cow can kick over the milk pail and I didn't want to get a bucket of milk all over me.

Gus asked me if I ever milked a cow. I replied that I had not. He drew my attention to the pail. "Lookie here," he said as I took a few steps closer and leaned down to see what was in the pail. Swish, swish, the warm milk hit me in the face and was dripping down onto my shirt. I got the next blast also. "Ha, ha ha!" Gus roared. "Hey mate, what do you think about that?" All of a sudden, Gus seeing the expression on my face said, "Enough of that, aye mate?" He finished milking Bessie, put a lid on the pail and set it by the door. Over on the wall he lifted a big leather harness off its pegs and carried it over to Dick's stall. The horses were in their straight stalls so he had to slide alongside the horse to put the harness on.

Gus draped it gently over the horse's back and made it fast. Next came the collar and bridle. The collar was a work of art, all bedecked with brass fittings and very shiny, especially the two finials on top. Then he harnessed Boy who was patiently waiting his turn but not forgetting the last of his hay. They stood quietly while I followed Gus out to the other barn. Gus propped open the door with a big iron bar. There inside was this beautiful machine with two wheels and a bar on its side held up with twine. This was some machine; I had never seen one like it before, not even in the city. I thought Brooklyn had everything, but this machine was so different. It had all kinds of levers and pedals. I wondered what it was and what Gus was going to do with it. The machine glistened; it was obvious that he took great care of it. Gus grabbed the shaft that stuck out from the front, lifted it up and walked out of the barn pulling the machine. I could tell it was heavy because he was huffing and puffing after he set it down. "Now," he said, "You wait here, mate. I gotta bring out the horses." He came out leading Boy backing him out on one side of the shaft and hooking him with chains. Then out came Dick. He was hooked up on the other side of the shaft, to the whipple-tree. He then strung lines from their bridles back toward the machine, knotted them and hung them up on a lever. The horses stood quietly, although every once and awhile they would stomp their feet on the ground probably to chase off flies. There were plenty of them around the barnyard as we were not far from the manure pile.

Now, I waited and wondered what would happen next. Gus had an oil can hanging from a bent coat hanger under the seat of the machine and I questioned how anyone could last long on that seat. I remembered going to the Metropolitan Opera in the city and having my behind fall asleep and it not waking up for a long time. He quickly oiled all the moving parts and gears of the machine with the oil, hung it back under the seat, grabbed the lines and mounted this thing. The horses started to shuffle back and forth and Gus, now up in the seat, started to rock back and forth. He made this funny clicking sound to start the horses walking forward. As the machine moved, there were all kinds of clicking sounds I could hear; like being in a store that sold clocks. Gus was bouncing up and down in the seat. The whole contraption rocked and rolled back and forth like a bobbing cork on a pond as he made his way to the edge of the field. I could see Gus pull back the lever and the bar go down. That started clicking too as the blades shifted back and forth. This was his mowing machine. Then Gus set the bar down on the ground and began cutting his first swath. The grass fell over the sideboard like falling dominoes. The mower worked its way to the end of the field. At the end of the swath, Gus got off the machine, retrieved his oil can and oiled the gears again. He climbed up again, backed the horses up and the machine mowed square the corner of the field. Each swath was meticulous; there wasn't a blade of grass standing. I watched for a while then headed on home I could see this would take all day. It was almost lunchtime and I knew Mom would be looking for me at the family table.

After lunch I hung around the house and did any jobs that needed to be done. Later, after we had dinner, I asked Mom if I could go back up to Gus's. She asked me again if I thought I was being an annoyance. I assured her that Gus was expecting me. It was a beautiful evening as I walked up the road toward the sunset. The heat waves shimmered off the road, rising as the cool of the evening set in. I knocked on the screen door as I always would do (part of Mom's upbringing rules.) "Come on in," Skipper called out as if expecting me. I walked in holding the screen door so it wouldn't slam as that always bothered Gus: that was his 'slam bam.' I took my usual seat, "Where's Gus?" I asked.

"He'll be out in a minute; he's shaving." Skipper answered.

"Hey mate!" Gus piped up as he stepped out from his room while brushing shaving cream on his face. In one hand he held a shaving mug, in the other a small brush. Gus was wearing a sleeveless undershirt and in the glimmer of the kerosene light, I could see his tattoos. Some were on each of his arms and one big one on his chest that I could just see over the top of his collar. I could make out flags and what looked like a ship's masts. "I'll be done in a minute," he said, and stepped back into his room. Skipper offered me dinner, saying, "Gus is having his favorite tonight, how about it."

"No thank you, I just ate." (I meant 'et) In a few minutes, Gus came out and sat down in his favorite chair. I couldn't help but notice that his chair was strung with rope and looked like it had been repaired many times. His clean-shaven face gentled his appearance. I was used to seeing him with stubble on his face. The fragrance of his shaving cream blended in with the smell of wood smoke and kerosene.

"Red at night, sailors' delight; tomorrow will be a good day for working hay," Gus muttered as Skipper brought over his dinner and set it down in front of him. The biggest plate of beans I have ever seen—beans from rim to rim. He mumbled, "Thank you, God," and went at it. Quickly, they were all gone. He wiped his plate with his bread, let out a burp, excused himself and crossed his knife and fork on this plate saying "Always ship your oars when you're finished; that way everybody knows you're all done." His knife and fork were crossed over the plate like a pair of oars in a row boat.

"Now, tomorrow, I have to turn the hay and I'd like it if you could help out. What do you say, mate?" I jumped right on this opportunity, "I'll be here, what time?" "Whenever you get here would be good, mate."

Leaning back in his chair, Gus looked over at me and asked if I knew anything about knots. Cautiously, I said "Yes". When I was a Boy Scout, I earned a merit badge in ropes but you never knew with a Gus question what he had in mind. Gus reached under his chair and pulled up a piece of rope. Just a coincidence mind you, I thought. He took the rope, made a few twists and asked me, "What's this, mate?" I recognized that knot, "That's a half-hitch," I said. "Very good, mate now how about this one?" as he twisted one knot after another. I couldn't recognize any of them. He just flicked his wrist, gave a twist, and out came a different knot each time. He called out each knot by name.

"Gus, what kinds of knots did you use aboard ship?" I asked. I was getting dizzy watching him spin the rope. "If you want to learn about knots and ropes, I'll teach ya, but it takes lots of practice," Gus proudly offered.

He began, "Ropes are used aboard ship for almost every purpose in sailing. Without ropes, you could not sail the ship. They are used for hoisting sails, making shrouds, ratlines, cables, hawsers and many other things necessary to operate a ship. Let's begin with this, do you know what rope is made of?" I shrugged my shoulders and gave him a puzzled looked because that was a "do you know?" question. "The best rope is made from manila fiber which comes from the leaf stalks of textile bananas from the Philippine Islands." Out came the atlas and "do you know where the Philippine Islands are?" he began inquiring. I leaned over the atlas, combed through the pages as if I knew where I was going, but it wasn't where Gus expected, of course.

"Good try. What do they teach you in school? Don't you have

geography lessons?" I just let that one slide by, picked up the rope and started fiddling with it. Gus grabbed it out of my hand, looked up at the ceiling and continued. "Manila fiber is very strong but it is hard to bend. When the weather is hot and dry, we have to wet it down to use. Another kind of rope is made from hemp and the best comes from Russia." His stubby finger went sliding along the map in the atlas obliterating half of my view of Russia. "It's very strong but not as strong as manila; that's why we tar it which strengthens it. Usually we use this rope for our standing rigging because it stays taut. There are times when we use rope made of coir which comes from the husks of coconuts. It makes the best tow ropes because it floats easily. The husks are woven into strands. A rope of three strands is called a hawser; a rope of four strands is a shroud-laid. The larger ropes are made by twisting three hawsers together which we call a cable."

Gus continued, "Small stuff are ropes used onboard ship for ratline which we use for rattling down the riggings and the rope ladders that go up the shrouds. A sailor never goes aloft without a supply of rope yarn just in case he finds a worn out housing or stop on the sail which always needs to be repaired right away." Gus got out of his chair, went over to his little bureau and brought back a metal spike. It was about a foot long, pretty thick with a hole at the top. Gus handed it to me; it was heavy. "We use this marline spike for repairing ropes, separating strands or joining two strands of rope together." Curiously I asked, "What's the hole for?" Gus looked at me squinting said "Oh, that's where a lanyard goes; to wear around your neck so you can have your hands free when you're not using it." Gus then retrieved his rope. "Make me a square knot, mate."

I took the rope, one end in each hand, crossed the ends over as I tried to remember the next step but somehow not getting the second twist where I wanted. It didn't look right. Sure enough it wasn't right as Gus grabbed the rope out of my hand, his patience wearing very thin. "Let me see that, mate!" as he fumbled with the ends of the rope, a twist here, a twist there and bingo he handed me the knot. "How about a bowline" he asked. "What?" my eyes darting back and forth searching for the meaning. He took the rope and flicked it around and there it was, a beautifully shaped knot. With each twist of his wrist a new knot appeared. "Back to the facts, a square knot is also a reef knot used to tie two ropes together," another flick and a twist, "a sheet bend mate, for tying two different size ropes together. Here now, the bowline, you use this knot to hoist a man aloft or for bending two hawsers together." Then came the most interesting knot of all; Gus's hands twisting and turning and then "Lookie here mate! A sheep shank. Ya use this to make a rope shorter." Gus laid this knot on the table. It was beautiful with three loops and when he pulled on its ends it didn't come apart.

"An important thing to learn, mate is that ropes and knots are the key to most activities that are done aboard ship. To be an able-bodied seaman ya just got to know all about how they work. So set yourself back and listen up. First there is splicing; that's where you put two pieces of rope together to make it longer. You know mate, there's no hardware stores out at sea. There are several types of splices. The first is called short splice. Then there is the long splice and the eye splice. The idea here is to join the ropes smoothly and strong, by first thinning the rope out and tapering it . That is what the marlinspike is used for. It is always used by the right hand; while the left manipulates the rope like so."

Looking through his 'Mariners Handbook', Gus turned to a picture demonstrating how the marlinspike is used by pushing the sharp end through the strands to separate them and then twisting the rope to work it.

"When working with a short strand, open the strands individually for a good distance to work. Then place the two ends together and bind it with twine and then braid the other strands together." Gus turned in his handbook to another illustration showing the different splices.

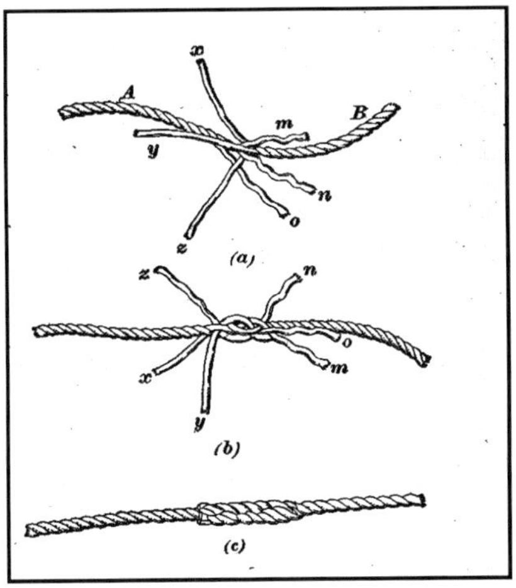

“Pay attention now, lad, watch how this is done. Place the two ends together so that each of the strands overlaps the other. Hold the other strands x, y, and z and the rope a here in the other hand, then take one strand like n and pass it over strand y and with an opening you made with your marlinspike push strand n through under x and pull it tight. Do the same with the other strands m and o passing each over the adjoining next strand and under the next one. Do exactly the same with the strands of the other rope by passing each strand over the adjoining one and under the next, like in this figure b in my book. Everything has to be pulled taut then you can cut off the ends of the strands. The long splice is similar but the strands are much longer. I don’t expect ya to know all this, just to get the idea that ropes are made to suit your purpose. Here is a picture of an eye splice and then the chain splice. Ya can see how they would be used. Can’t ya?”

Eye splice

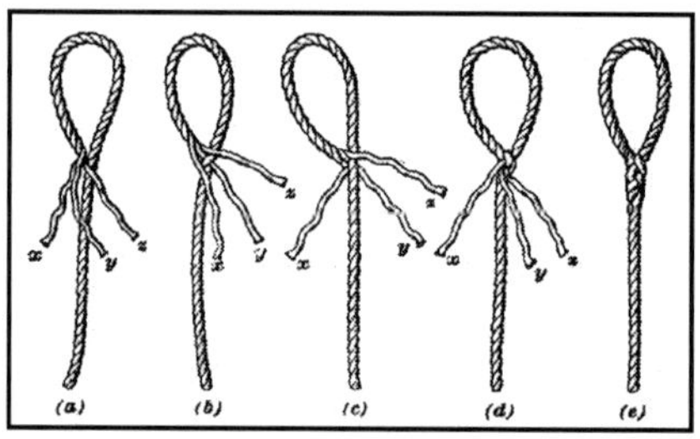

Chain splice

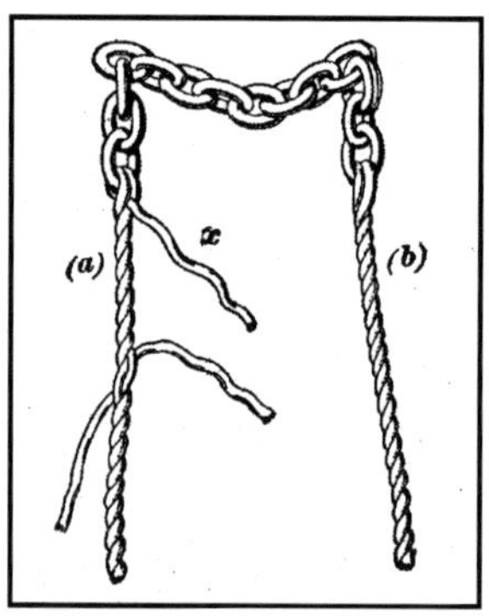

"Ya can see it's more than just tying a knot around something. Out there at sea rope has to be strong. Can't take any chances for slippage or weakness; it could cost ya your life. Here is a diagram of many of the knots we used and their use with the tackle and blocks we used for hoisting or making fast."

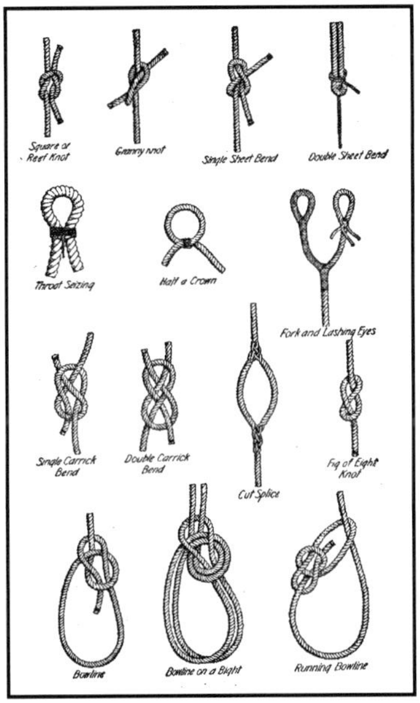

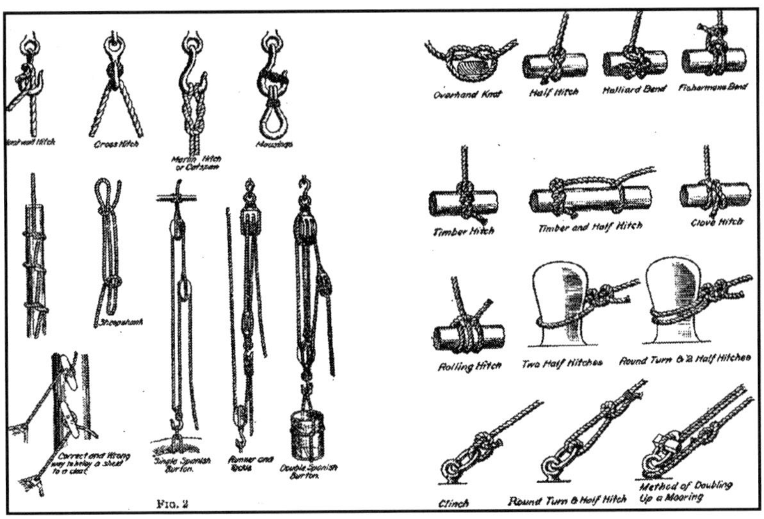

FIG. 2

"Well, mate, tomorrow's another day. Gotta get up early to turn the hay. If you want, you can give me a hand."

Gus tossed over the rope at me saying "Take this home with you and practice the knots I just taught you." "Goodnight, Gus. Goodnight, Skipper" and off I went, shuffling down the road, swinging my rope round and round in the air like a cowpuncher.

The next morning came fast and I got up with the sun. I had some cereal and then out the door I went. Mom saw me leave and yelled out the door behind me, "Make sure you're home for lunch!" "Okay, Mom," I responded as I went high stepping up the road.

Gus was in the barnyard with another machine he just dragged out. This one also had two wheels, like the mowing machine, but they were much taller and between the wheels were long curved metal spikes with sharp points that looked like a metal rake.

Gus's hay rake as it appears today

Boy was harnessed as Gus backed him between two wooden shafts then hitched leather lines he called traces to a whipple-tree. This was Boy's job and he was anxious to get going, swishing his tail and stomping on the ground. Gus looked over at me and said, “If ya wonderin’ what this is mate, it’s a hay rake and we’re goin’ to turn the hay over to dry.” Gus climbed up on the metal seat, another rickety contraption I thought. The seat bounced up and down every time he moved, “mate, bring along the wooden rake layin’ there by the barn.”

The hay rake which was ten feet wide took up the whole road. This machine also had its own musical sound, the metallic ringing of the wheels as it rolled over rocks, the clicking of the gears each time the rake was raised and lowered.

With Gus’s customary click of his tongue, Boy ambled down the road toward the hay field. The hay rake clacked and swayed from side to side, Gus perched up on top with nothing really to hold on to except the drop rake lever. Cocking his head around, he yelled out “mate, you rake up the loose stuff into the windrows, ya got it?”

I walked behind the rake doing my job, looking up every once in awhile to see rows of hay being made into windrows and the teeth on the rake rising up and down. Down below, I saw lots of creatures scampering for cover—grasshoppers, crickets, and even a few smaller grass snakes. Frogs were jumping right and left to get out of the way after being roused from their business. Birds started flying in: robins, cow birds and even hawks appeared overhead. They began swooping in to peck away at these vulnerable creatures. Up and down the field I followed the hay rake as Gus turned the hay into neatly crisscrossed windrows to dry in the sun.

After Gus had raked almost half the field, he pulled the horse up and waved me to come near. I came alongside, keeping clear of the horse’s tail swishing away the deer flies that swarmed about whenever you stopped moving.

"Do you see how I'm doin' this?" Gus leaned forward and asked, lines in hand climbing down from his seat. "Okay mate, get up on here," handing me the lines. "See here mate, step on this peddle and the rake will come up. Push this handle forward and the rake will drop. Line up the rows as I have 'em, okay, mate?" He paused, "Now be off with ya."

I was excited. Gus was going to let me drive the horse and rake the hay. "Hold the lines in your left hand," he instructed as I climbed up onto the seat. I laid the lines in my left hand with my right hand on the lever for support. Gus gave a click and off the horse went. I felt the rake swaying, a little to the right and little to the left; I guess because it was so wide. I approached the next swath and pushed the pedal. The rake came down with a clatter as the teeth bounced across the ground rolling the hay in its path. When I came to the windrow, I pulled on the lever and the rake came up dumping the hay right alongside. Boy just eased along. He probably could have done this job himself. When I said "Whoa," he stopped and when I "clicked," he walked forward. Up and down the field we went, Boy swishing his tail back and forth, the gentle pull of the horse, and his smooth even gate. As the hours passed, I finally did up the last windrow and then headed in. A gentle pressure on one line and we turned and headed toward the barn where Gus was waiting. I got off and set the lines on the handle just like I had seen Gus do. "Thanks a lot, Gus. That was exciting and fun." Gus unhitched Boy, put him in his stall, un-harnessed him and rewarded him with a bucket of oats and a pail of cold water.

Dad would be arriving from the city for the weekend and I couldn't wait to tell him about driving the horse and hay rake. I also knew that I had some math problems he left me and some explaining to do about the ones I didn't finish. The plan was to go home for supper, handle a few chores and then I'd be back up to Gus's. Evening came, I did the dishes and out the door I went. I wondered what we would talk about tonight, maybe some more stories about sailing ships. I knew also I couldn't stay too late because Dad would be coming up later. I found Gus at the table with the kerosene light turned up high, so high that trails of think black smoke were curling above the lamp chimney. All about the table were small pieces of wood; in Gus's hand a little model boat that he was whittling on with a pen knife and in front of him was a quart wine bottle.

"Hi Gus, what are you doing?" I asked curiously, pulling out my chair and sitting down.

"I'm making a ship in the bottle," he mumbled, obviously preoccupied. Spools of thread, little jars of paint and a ball of clay were on the corner of the table. "How are you going to get that little ship in the bottle," I pressed on. "It's a secret," he blurted out, his eyes squinting as he started to insert little threads into the model's mast. Inside the bottle he had clay pressed on the side with little wooden houses and beneath, clay molded and painted like waves with little white caps. Gus worked away and I watched. He obviously was not going to reveal any information about his 'secret.' I just couldn't imagine how he was going to get that ship with its mast and rigging through the neck of the bottle. One thing I did notice was that the mast wasn't glued because he kept moving it back and forth on the hull as he adjusted the rigging. He held the little ship in one hand, pulled on the thread with the other. When he did that I could see the masts could lay flat and when he pulled the threads the masts came up again. That must be the secret, I figured.

Skipper put on some tea and we spent the evening watching Gus work on his ship in the bottle. Deep down inside I thought I may have found the secret but never having tried it, I still am not sure how he really did it. “Where did you learn that?” I couldn’t resist. “Aha,” Gus looked up, paused and then admitted, “I learned how from an old sailor but don’t bother to ask anything more because like I told ya, it’s a secret.” Gus put the little ship with the masts down through the neck of the bottle and when in place, he pulled on the threads and up the masts came, a ship in the bottle. That ended this discussion; a secret must be a secret. For a man with such big hands and stubby fingers and, remembering one part of his finger was missing, it was amazing to watch how he could carve out such minute detail, even threading the masts with sewing thread. It all looked too painstakingly difficult in the glow of a kerosene lamp. It was getting late and Gus picked up his tools, put the ship in the bottle on the sideboard, swept up his mess of wood scrapings and tossed them into the stove. He leaned back in his chair and planted his arm on the table, opened that big mitt of his and stared me square in the eye. “Let’s go mate, grab a hold.” I turned my chair so I could get my best grip and the leverage I would need to keep my hand from slamming on the table. That’s what I thought anyway. I felt my hand sink into his and there was that stub of a finger to deal with as usual. Back and forth we went. Every once in while, I’d get his hand to move a little then he would start to distract me and make me look at him and when I did his eyes would squint, then his stare and bango, down I’d go. “Gus, I want to ask you a question, if you don’t mind,” and as I was asking I could hear the phone ringing. Skipper had a party line with different rings. This call was for Skipper. I hoped it wasn’t my mother; it was getting late. Sure enough, time to head home. Dad must be there so down the road I went, running all the way. As I walked in, Mom was by the sink and Dad was sitting at the table. "Hi Mom,. Hi Dad," I said and gave Dad a big hug. Mom looked at me with a serious expression on her face and said "I think you're spending too much time up the road. We have things to do down here. I think you should be spending more time at home."

Well, I knew I would have to figure out some arrangement that would be satisfactory with Mom. I didn't want to miss any of Gus's stories yet I didn't want to upset Mom. If I stayed home in the morning, then spent the afternoon and evening up at Gus's, maybe that could work. I started spending a lot of time helping Mom. I weeded her garden, vacuumed the floors, mowed the lawn. I did something every day. I felt better because Mom seemed happier with me now that I was doing some of the things she needed me to do. I spent more time with my brothers and sisters, blazing trails and exploring. My sisters were too young to be included in everything, especially going in the woods. Mom wouldn't let them anyway.

After a couple of days or so of not going up the road at all, I asked Mom if our arrangement could start now. She agreed. That afternoon I went up the road to Gus's. In the south meadow, Bessie was lying down in the hay field taking in the sun or waiting for a rainstorm. I wasn't sure—it could be a thunderstorm coming in. Gus was sitting on the wellhead with a broom handle stiff-armed, holding it out in front of him. Hanging from a rope tied to the broom was an old steam iron and he was curling it up and down. Up and down the iron went with his stiff arms out twisting round the handle. I sat down next to him and watched this extraordinary feat. Gus's forearm was bulging, tattoos wiggling under the strain of the iron's weight.

'Bessie' may be expecting rain?

"Try this, mate." He handed me the broom handle. "Hold it away from you, stiffen your arms out and twist the broom with your wrists." It was easier said than done. Winding the rope around the handle was the easiest part but once the slack was out of the rope, the weight of the iron took hold. I had all I could do to keep the iron from pulling my stiffened arms down.

I lasted for about one minute and that was the end of this activity. Gus took the rope off the handle, stood up and, pointing the tip of the handle into the ground, holding on with both hands, he started doing twists under and around the broom handle. Upside down, inside out, I thought this must be some kind of exercise sailors did onboard to keep up their agility.

Gus set the broom handle down, opened his hands, spit in his palms and then squeezed his hands together. Out came this loud swishing sound like someone blowing in a bottle. In and out he pressed a rhythmic beat, swoosh, swish, swoosh, swish and then out came a chanty. "Rollin' home, rollin' home, rollin' home across the sea, rollin' home to dear old England, rollin' home dear land to thee," singing and tapping his feet.

"What do you think of that, mate; do you like that chanty?"

"Yes, that was great. Were you a chanty-man?" I asked. "That's an old chanty we used to sing onboard when we were sailing back to England after a long voyage," Gus replied.

"Gus," my curiosity further aroused, "how did you learn these songs; do you know any more?" "Well," Gus said, "I learned them from all the different crews I worked with over the years. Chanties play a very important role onboard ship. When there are different jobs to do, there are special chanties we sing. To mast head a topsail yard or any hauling, a long drag chanty is sung. To haul on a taut rope or to sweat up a halyard, we sing a short drag chanty. When we have to pump the ship, we use a tail rope which we put on the pump handles then the chanty-man may sing a pumping chanty. We old salts sang all the chanties. Do you want to hear another? This here is a long drag chanty and it goes like this." Pressing his hands together and making that swishing rhythm he piped up, "A Yankee ship came up the river, blow boys blow, a Yankee ship with a Yankee skipper, blow my bully boys, blow."

Gus's song rang out through the barnyard; he really got into it and suddenly stood up and started to do this dance. Rocking back and forth, his feet tapping on the stone walk and spinning round and round, he called it a sailors' jig. "Gus," I asked "Can you teach me how to do the jig?" He laughed and looking at me, said, "We'll do more later." Just then Skipper came out the door laughing to herself and offered us a glass of fresh squeezed lemonade. We all sat on the wellhead, gazed out into the west meadow and watched the sun set behind the tall maples. The air was cooling off and we could see a trio of geese making their way back to the swamp bottom. Peepers started piping up their evening song from down by the pond. All was well 'up the road.'

Skipper told me about her ship named the *'Nebula'*. "She was a cutter, two-masted, fifty-five feet long and we competed in races and did charter work," she said. "We sailed everywhere, South America, Mexico and all along the Caribbean. She was a good ship, isn't that right, Gus?" He nodded and added. "We weathered some pretty good storms and could always count on her; a fine ship, she was."

"What happened to her?" I asked inquisitively. Gus explained, "She needed a lot of work because she was already sixty-five years old and no longer seaworthy." Skipper appeared sad as she related the good times aboard the

'Nebula'. She pointed toward the screen door and the bell that hung alongside. It was shiny and had a leather thong hanging from the clapper. "That's the bell we had on board the *'Nebula'*."

Gus and Skipper spent the rest of the evening talking about old times as I listened intently. We went inside as the air was cooling off and Skipper put up a pot of tea which we all enjoyed.

It was getting late and I knew I had better be on my way. As I began to leave, Gus reminded me, "Mate, tomorrow can you help me get the hay in?" he asked. "That's right," I remembered, "Sure, Gus." I said my goodnights and off I went.

When I arrived home, Mom was in the kitchen as I expected. "Mom, Gus needs me to help him get the hay in. Can I go? Please get me up early," I pleaded. Morning came and the sun struggled between the clouds as I headed up the road. Did those clouds mean rain? I wasn't sure. Gus was in the barn harnessing the horses and out in front was the hay wagon, ready to go. Gus handed me a three-pronged fork and helped me up on to the tail of the wagon. Each horse was hitched and ready to go. "Come on, Dick, come on Boy!" With a click of his tongue the wagon started to roll. I almost fell off the back from the jerk of the horses' first pull. Creaking along, the wheels clanging on the stones we headed out toward the freshly mown meadow. Gus drove the horses between the windrows. They were spaced just wide enough for the wagon to fit between. It was a long flat wagon with wooden wheels. Looking at it, I wondered just how it could hold a lot of hay.

As we pulled up to the first windrow, the horses stopped. I slid off the back of the wagon with my fork in hand, ready to load the hay. Gus piped up,

"Now lookie here, this is how it's done," he said, as he took the fork and drove it into the rolled hay. "Just put your fork in the hay, lean forward and give a twist like so," and then in a lifting rolling motion he placed the hay in the front of the wagon.

It looked easy enough, but it took me a while to get the hang of it. The wagon was loaded from front to back and each pitch overlapped the one before. Gus was very particular about where we put each pitch of hay. The horses moved forward and stopped with each command. The hay started to get higher. Now Gus was atop the load and I was pitching up the hay as high as I could. Gus grabbed my pitch with his fork and placed it on the load. Higher and higher Gus rose as more hay was put on. He had the lines attached to a pole and kept adjusting them further up. As he bounced back and forth across the load, I could see it was getting pretty top heavy. "The clouds are getting darker and the air feels like rain, heave away," Gus bellowed down.

Finally, we got to the last windrow. Gus made some adjustments to the load, his legs now sinking deeper and deeper into the hay as he made his way forward to where the lines were. "Come on, Boy, come on, Dick!" Let's get her in," Gus piped down. I followed behind with my fork on my shoulder. The horses moved ever so slowly; the wagon swaying under the load. I thought for sure every time the wagon wheels lifted over a rock, the load would tip over. Looking up, I could see Gus; lines in one hand, hanging on the pole with the other as if he was riding out a storm.

Gus drove the wagon round back behind the barn where the hay mow was and drove in. There were two mows, one on each side of the wagon. How clever, I thought; all we needed to do was pitch the hay off, right and left. The hay was now in and one more job done for winter.

We put off unloading the hay until the next day. Gus unhitched the horses, walked them forward into the main barn, un-harnessed them and rewarded each of them for their hard work with oats and water. The sky was darkening fast, rumbles of thunder could now be heard getting closer. We headed for the house and I thought I should get right home because Mom didn't like thunderstorms. Down the road I ran as flashes of lightning now filled the sky.

Out of breath, I made my way into the kitchen. Mom and the kids were huddled up on the couch in the living room. I sat at the table and waited for the next flash. Lightning storms up here on the hill are spectacular. Flashes are brilliant and one can easily see the forks of electricity strike the ground. The whole sky lights up, the thunder shakes the ground and sometimes it's so loud, it's very scary. Then I counted—one, two, three and so on until I heard the clap of thunder—each second equaling about one fifth of a mile. The total number of seconds divided by five or so gives the distance in miles of how far off the lightning is striking and if it was getting closer or moving away. That is one of many skills Gus taught me.

A lightning flash followed immediately by a loud click meant a very near or direct strike — either to the house or a tree in the yard. Luckily, that didn't happen as the storm passed over quickly. We all went to bed thankful that we didn't have to stay up all night listening to claps of thunder and watching the room light up with each bolt.

My arms and legs suddenly felt sore from all the hay I heaved. I must have pitched a ton of hay up on the wagon that day. It didn't take long to fall asleep.

The next morning while I was mowing the lawn or at least trying to with our old hand reel mower, I saw Gus walking down the road carrying a black lunch pail like the one I take to school when I'm back home. "Where are you going, Gus?" I yelled out. "To work, mate," he replied. I ran up to him and together, we walked down the road "Where do you work?" I asked.

"Oh, down at the mill, mate," he mumbled. I walked down the dirt road to the highway with him. It was hard to keep up with him as he walked so fast. Maybe he was late, I thought. It was a good mile and a half to the bottom of the hill. We walked through swarms of deer flies that seemed to hang around the corners of the road. At the bottom of the hill, we sat on the old milk platform. Gus told me it was used by the old farmers. They would bring the milk down by horse and wagon and set it on the platform to be picked-up and put into a tanker. A car approached and slowed down. It was Gus's ride. "See you later, mate," he said as he got in. "Goodbye Gus," I waved.

It was a lot harder walking up the road, especially up the big hill. It had two turns on it. Gus told me the road was built that way so, in the old days, when horses pulled up wagons, they would stop at the turns and have a chance to rest before they had to begin the climb again. So when I got to the turns, I did the same to catch my breath. The backs of my calves ached as I headed out across the flats. Around the first turn, a swarm of deer flies waited for me on the way back. These deer flies were very persistent and would circle around my head. Very annoying. I figured out a way to get them by raising both hands above my head, palms facing one another and when the fly would inevitably land on one hand and at the right moment when he wasn't looking, I'd clap my hands together, bango, he's done for. It was a nice quiet walk up the road. I can't tell you how many times over the years I saw a partridge or a fox cross the road in front of me.

I spent the rest of the day working around our house. Mom had a list of chores written out and right on top was my name. After an early dinner and while finishing the dishes, I saw Gus walking back up the road, swinging his lunch pail. I ran out the door to greet him, "How was work?" I asked, as we walked together up the road. "Can I come up after dinner?" figuring maybe he was tired from work, like Dad gets when he comes home, plants himself on the couch and can't talk to anyone. Gus answered "Sure, mate, see you then."

I waited for a half hour to pass then started up the road. Gus was at the well pumping water into a bucket. There was no running water at his house and no electricity, just an old 'Aladdin' kerosene lantern. Gus carried in the water, set it down and took his chair. "Hi, Skipper, how's everything?" I asked. She was washing dishes in a copper kettle on the stove. Gus stretched his arms then reached for his java, as he commonly referred to his coffee.

At the ready and sitting on the table was his blue atlas. He flipped through the pages until he found what he was looking for. I leaned over trying to follow but couldn't see much because his finger blocked the page. He stopped somewhere below England. That I did know from geography in school and I piped right up. "There's England," trying to make a few points as Gus was always nailing me on my geography. "There's Liverpool," he said, pointing with this finger.

Gus slid his finger down around the tip of Africa then east through the Indian Ocean and down across the Pacific Ocean toward the west coast of the United States. "Here's San Francisco. Do you know how far that is from Liverpool, mate?" he queried me. "Looks like a real long way," I answered. "That's right, mate. It's about 14,000 nautical miles and it takes about a month to make this voyage. This was the sailing route before the Panama Canal was built in 1914."

"Down around Cape Hope, at the southern tip of Africa, that's where the Atlantic and Indian oceans meet. The waves and the currents can be very treacherous. The waves are pretty high in those parts. All hands are busy when they sail around the Cape."

"Up aloft was especially tricky, sometimes you have to hang on to a line for dear life. The captain would be watching the barometer very closely and if it rose or fell fast, you'd know there was a storm up ahead. The barometer usually rises for dry or wet weather, winds from a northerly direction or from a decrease in wind. It stays pretty much the same in the tropics because the winds generally blow in the same direction."

Gus, now circling with his finger, showed me the direction wind currents can take in the northern and southern hemispheres. "When bad storms do appear, the barometer falls very low before approaching the center of the storm but once through the center, it can rise quickly back to normal. On some coasts, the barometer is affected by the wind depending on whether it blows from sea or land. The mercury can rise on the approach of a sea breeze and to the setting in of a land breeze." This was absolutely the most interesting subject, I thought. I couldn't help but wonder how sailors could tell what weather was up ahead by using the few instruments they had. Imagine how it was when the earliest ships ventured into the unknown oceans.

Gus continued, "These, mate, are the general rules when reading your barometer: a gradual rise means unsettled weather, a rise with dry air and temperature cooling in the summer means winds from the north, but in southern latitudes, it means winds from the south. If rain falls, it means better weather coming. A rise in the barometer with moist air and low temperatures means wind and rain from the north in northern latitudes and from the south in southern latitudes."

If it wasn't for Gus's finger going 'round and 'round and his hand gestures, this would be hard to remember. How could I forget? During our relationship, Gus would go over this a hundred times with me. "To further complicate matters," Gus went on, "a rapidly falling barometer always means stormy weather which if coupled with a northerly wind, can be an indication of rain in the summer or snow in the winter."

"Another way to read the weather is by looking toward the sky and its colors. A red sky at sunset means fine weather but a red sky in the morning means bad weather, windy or rainy. A gray sky in the morning can usually mean okay weather. Soft and delicate clouds foretell fine weather with light breezes. Hard-edged or oily looking clouds can mean windy. If the sky is a dark and gloomy blue sky, wind is likely and a light blue sky means fine weather."

"Usually, the softer the clouds, the lesser the wind, the harder tufted or ragged the clouds, the stronger the wind will be. Just let me add one more: a bright yellow sky at sunset may mean wind and a pale yellow sky generally means wet weather ahead. Do ya get all that, mate? Don't worry, you will before I'm through with you."

I sat back and thought a minute. "How can anyone remember all these things? I asked myself." "What happens when you're sailing and a big storm is coming up? What do you do?" I asked.

"Well, mate, the first thing you have to do is to find out where the storm's center is and then plot a course around it. This is how you do it. First you must face the ship into the wind and from there you find the center of the storm. If you're in northern latitudes, the center should be ten or so points to your right and if you're in southern latitudes, you mark the center to your left. If you find yourself in the middle of a storm and your barometer is falling fast and steady at about one-half inch of mercury, the bearing of the center of the storm may be at eight points. Knowing the bearing, you need to find your ship's heading, your position in relation to the track of the storm. You do this by observing the shift of the wind. In the northern latitudes if the wind shifts to the right you are then to the right of the storm track or if the wind is shifting to the left, you are to the left of the storm track."

Gus continued, "If the wind is steadily shifting and slightly increasing in speed that means your ship is on the storm track and most likely you are in front of the approaching storm. When you are well within the storm center, the barometer will drop fast, the wind will pick up and there will be heavy squalls. There could be lightning, rain, heavy seas and lots of shifting waters. As the storm passes, the barometer rises up again, the winds decrease and the weather will start to clear but you have to be mindful that the seas continue to be very dangerous."

"You asked me, what do you do? The captain has to maneuver his ship so as to stay away from the storm center as best he can by bringing the wind on the starboard bow, keep closed-hauled on the starboard tack and making as much way as he needs to make every effort to avoid the center of the storm. The best way keeping in mind that storms can curve to the northeast."

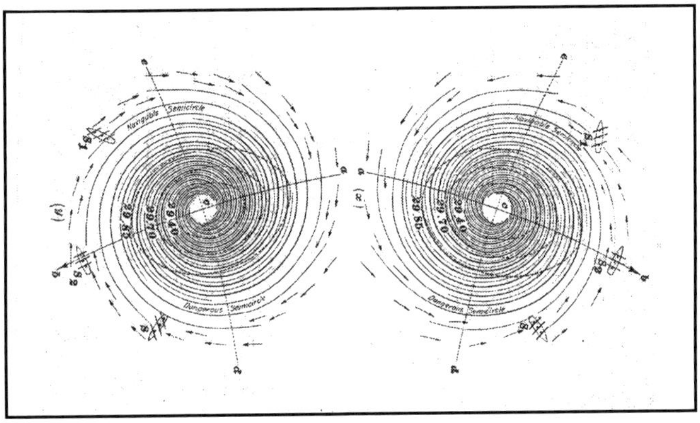

Storm navigation Techniques

"Did you get all that?" Gus said chuckling. "Here," he pulled out a diagram he had tucked in the back of his atlas. "This will show you; take a look here, mate." Pointing to a diagram, Gus further added that, "Cyclonic storms are very apt to be followed by another storm, so, if you escape from one storm center, you could find yourself involved in another that even may follow the first storm track. Typhoons in the western Pacific Ocean are very much the same as West Indian hurricanes of the Atlantic but storms coming up from the tropics have the tendency to re-curve especially in the higher latitudes."

"Now while we are on the subject we should talk about the compass and its use, this is what guides our ships today. In the old days it was the stars but there was no guarantee that a ship could get to the same place twice. There is a bit to know about this so pay good attention. This is important. First, the earth is a huge magnet having magnetic poles in the vicinity of the geographic poles. Ya know, the North and South Poles. The needle points to the North and the other end of the needle points south. Sometimes this is called the red pole and the blue pole because of its magnetic polarity. Now on board ship especially metal ships there is the chance that the needle is not quite true. This is called deviation, or deflection from the magnetic pole or meridian. So this is an error that must be corrected and taken into account when plotting a course. The major principle in compensating for a compass is to counteract the effect these magnetic disturbances have. This was usually done by placing little magnets around the compass to adjust for the deviation. If the correct magnetic bearing lies to the right of the compass bearing, then the deviation is easterly. If it is to the left, then the deviation is westerly. There are deviation tables that are used for this adjustment. Sometimes new deviation tables have to be made because different cargos may have a magnetic quality that will also distort a compass. This is an important job for the navigator in monitoring the performance of the compass. Even the vibrations from an engine propeller can make the needle on the compass sluggish and the navigator would have to recharge or re-magnetize the needle. That's why compasses should always be located away from electric motors and electric wires."

"Now let's get on with plotting a course or at least try to. A compass course is the direction or course steered by a ship which is affected by variation, deviation and other factors. A true course is the compass course and the corrections made for deviation and leeway. What is leeway? This is the result of the pressure of the sea and wind that is exerted on the hull and sails which causes her to drift sideways and is subject to the amount of wind and the shape of the hull. This is expressed in points. To find the true course the navigator has to apply variation and deviation by allowing leeway in the direction in which the wind is blowing."

"The compass card represents the visible horizon. The person using it is considered to be in the center of the card which means that when you make corrections you need to imagine yourself in the center position. Names of points and number of degrees and seconds are found on a table the navigator keeps for this purpose. This card is divided into degrees from 0 to 360 beginning at the north point."

"There is a rule that is applied when using this card. If the compass error is east, the true course is greater than the compass course; if it is west the true compass course is less than the compass course. There are special arithmetic tables to precisely determine your true bearing. Some day maybe you may understand this, after all it took sailors hundreds of years to get to this point."

"Well anyway, all this work is called 'the day's work' whereby each noon the course and the distance made during the past 24 hours are plotted. Each compass course is converted to true course, the longitude and latitude, distance and then all plotted on a map. The official log book contains this carefully put together record of the day's work to include all important things that may have happened during that time. This is recorded at the end of each watch by the officer in charge of the deck which he enters in a 'scrap log' which is later put in the official ship's log."

Gus laughingly muttered, "We'll make a navigator out of you yet. The kinds of weather you have to really keep your eye on occur during periods of hot spells, variable winds, increasing humidity and changing barometric readings. Every strong wind or gale must be treated as a possible cyclone storm."

"Gus," I interrupted, "How did you learn all this?" Gus adjusting himself in his chair, reiterated, "Well, mate, I'll tell ya, when you've sailed on as many ships as I have and in all kinds of weather, you kinda' get a feel for it. It gets under your skin; you can smell it and actually feel it coming."

"Gus," I asked. "What can you tell me about navigation, like how does the skipper know how to get from one place to the other and where he is when in the middle of the ocean?"

"Well mate, I was only a bos'n. Navigation is really an officer's thing but I can tell a little something about it in order for you to get the idea."

"Well, we just talked about the compass being an important instrument in knowing where you're going but you have to have more information than that especially when you're in the middle of the ocean and all you see is water. There are other instruments we use as well. For instance a sextant is used for celestial observation and charts with longitude and navigation lines on it. To do this job right you really have to be good with arithmetic which to be perfectly honest with ya, I wasn't. But anyway I can tell you about one method called the Sumner line. You fix the position of your ship by astronomical sightings and then the intersection of lines plotted from that information. This line, the Sumner line is made by the right angles made from the observed citing be it a star or the sun."

"You can get this line whenever you can see the sun or spot a star or planet. I can tell about it but the arithmetic part you'll have to work on."

"When measuring the sun's altitude say in the morning you have to figure out the longitude and the true azimuth. These tables are found in the navigators book of tables. The azimuth is the sun's true bearing at the time the altitude is measured. The navigator then plots the position on his chart and using the longitude and latitude, he draws a line perpendicular to the sun's azimuth; it's true bearing. This is called the Sumner line. His ship is somewhere on that line as long as he didn't make any errors in his calculations. His exact position depends on how accurate the latitude is determined in order to find out where exactly he is on that line. He has to calculate a second line a short time later, so that the sun has changed its azimuth enough to be able to see a point of intersection between the lines. Since I told you the ship is on the first line somewhere, it's also going to be on the second line and hopefully exactly where they intersect. The first Sumner line is moved parallel to itself depending on the course and distance the ship has made over that time period. Then, using the second Sumner line, the intersecting point will be exactly in the position he took the second observation."

"This is a very useful navigational tool and can be used when navigating along a coastline. It can be used in combination with chain soundings or the bearing of a fixed light or object. The best and most reliable Sumner line is obtained from a star at twilight when crossed with a line by the sun or a line from the sun in the afternoon crossed with a star at dusk or during the evening."

"If ya take a bearing using a compass you have to allow for the different deviations we spoke of before: the difference between magnetic and true north and the magnetic potential of the ship as well. Also let me add one more thing. If you ever by some chance become a navigator, always use a light lead pencil and when you use the dividers don't punch holes in the skippers charts."

It was getting late. The kerosene light was flickering, probably running low on oil. I had to be on my way. It was a very dark night so Gus handed me his flashlight. I don't know if that was a good thing because when I shed the light on the side of the road, I could see the reflections of creatures' eyes.

Down the road I went, this time trying a tip-toe tactic so maybe no critter would hear me and think I'm their evening snack. At the house, everyone was asleep; it was really late. It's a good thing I wasn't home in Brooklyn. First thing, Dad had a rule when we went out, he expected us to be home almost always at 10:30 p.m. On weekends, when I would go out to the movies or just hang out on Kings Highway with my classmates, time would soon pass the 10:30 mark and when I realized that I was late, I would hightail it home. At the house I could see Dad sitting at the dining room table drinking a cup of tea, one of the ten or more cups of teas he had daily. The dining room was right off the foyer by the front door so I would sit on the front steps and wait, sometimes for hours. Finally he would go back upstairs sometimes at 1:00 or 2:00 a.m. at which time I would make a bee line for my room. The next morning when I came downstairs, he would be sitting at the table drinking his morning tea and would look up and say, "And, what time did you pull in last night, Mister?" I would say about ten o'clock, hoping that he would have still been in bed before he came down for his night tea. It wasn't very honest, I admit, and it did eventually have its consequences when he later found out and I was grounded.

Gus, on the other hand, wouldn't let me get away with that the first time I told him a fish tale. If he asked me a question and had any doubts about my response, he'd call me on it. So I found out the hard way that it was better to be 'up front' right off as not telling the truth always resulted in more untruths and, of course, a mark against my character, an integral part of our relationship.

I slipped into bed and thought over what Gus had taught me about navigation and weather. I concluded that the more I repeated the facts to myself the better chance I'd have to remember them. I tried to retain as much as possible also because it was likely that Gus may quiz me and I was expected to remember all he taught me. Later in my life I did become more and more familiar with reading the signs and being fairly accurate at predicting the weather, after all that was the way it was done back in those days. Imagine how the early farmers knew when it was going to be a good time to do the hay; wet hay is no good.

3

Shipping My Oars

The next morning we were all busy loading up the car with bags of clothing, souvenirs from the woods and even a few creatures handily collected in glass jars. We were ready to take the five hour journey home. Everyone wanted a seat by the window or up on the back window deck. Finally we were all settled, any disagreements ironed out by Mom. Everybody would have their chance. Off we went down the road, onto the highway, heading back home to Brooklyn.

My mom always made it a point to pack a little snack for Dad, a few pears, a couple of plums and maybe some grapes she put in a brown bag to keep on the front seat. After four or so hours, we were in the city and when Dad had just finished eating a pear he inadvertently threw it out of the window. Believe me, he couldn't have known. Had he asked us if there was anybody behind us, I'm sure it wouldn't have happened, but needless to say, it certainly did. Within a few moments, a motorcycle drove up alongside and perched on top was a man in blue—the law. Our eyes were popping out of our heads, "Look at that, wow!" Look at that was right; he was frantically waving for Dad to pull over. The traffic was heavy; I think we were on the West Side Highway, down by the piers. Everyone was now slowing down and staring as Dad pulled off onto a breakdown lane. The motorcycle cop was in front of us as we both came to stop. The cop got off his 'cycle, kicked his stand and started sauntering back toward us while reaching in his pocket and pulling out some kind of black leather book. He was wearing high black shiny boots, his cap tilted to one side like a bomber pilot and a great paunch hanging over his gun belt.

At the window, the cop said to Dad, "Let me see your license and registration!" Dad fumbled in his pockets. He never carried a wallet so God only knew where it could be. Finally he found it and handed it out the window. The cop looked at the license and then at Dad and back again at the license, then at Dad again. Finally the silence broke, "So, you like fruit, do you?" With that, Mom's face turned brilliant red, and as expected, shifted her eyes back and forth and then out came that little huffy noise. We were all studying the facial expressions between them and waited breathlessly each next move.

It came from Dad, "Care for a plum, Officer, or how about a pear?" The cop looked at Dad and we could see steam rising from under his cap. Responding abruptly, "I've already had one," the cop's hand wiped something off his face, his grimace reflected his outright frustration with the event. Reaching in his shirt pocket, out came a pencil and his book flipped open. Licking the end of the pencil, he began writing. My brother David blurted out, "What's he doing, writing Dad a ticket? He can't do that, doesn't he know who Dad is!" The cop tore out a paper, flipped it in the window onto Dad's lap, gave us a stare in the back seat and walked back to his cycle, stopping briefly to give a hitch to his trousers. He jumped on his cycle, gave a kick and off he sped back onto the highway.

"Dad, Dad! What did you get a ticket for?" We all said in unison. Mom picked up the ticket as Dad, now furious, peeled off onto the highway, shaking his fist out the window, muttering, "*Ver geharget*" or something like that. Mom read the ticket and mumbled "Oh, Ben," then folded it up and put it in the glove compartment. "What's it for, Mom, what did he do?" we all wanted to know.

Well anyway, back to school and the same old classes, day in and day out—get up in the morning, rush out the door, run down the street to the schoolyard, stand in line and then march up to class. In order to occupy my time in class, I began the habit of drawing pictures of sail ships in the back of my loose-leaf notebook. I had a special place coded with its own tab just after the one marked, 'Math.' The teacher would, during the course of her lesson, walk up and down the aisles and as, she was about to approach me, I would flip to the appropriate section. When she had passed, I would flip back and continue with my drawings. I drew all kinds of ships. I was pretty good at it, too. If I wasn't doing that, I'd be looking out those big classroom windows that every New York City public school had and I would gaze at the skyline from my fifth floor classroom.

On clear days, I imagined I saw all the way up to the Berkshires. I was even nominated by my classmates to be the window monitor and would raise and lower the windows with a long pole whenever the teacher asked me to.

One morning while working in my loose leaf, flipping back and forth between the math problem page and my doodling section, there was a knock on the classroom door. I looked up, as did the whole class, and I saw my Mom and Dad being greeted by the teacher. "Cheese and rice," echoed in my mind.

"Thanks for coming," she said and ushered them in. I sat in my seat paralyzed, never imagining that my parents would ever come to school.

This must be a parent conference or something, I thought. Mom never mentioned this morning that she'd be coming to school with Dad to see the teacher.

I felt I always gave my parents good reports on how I was doing even though my report cards didn't always agree. My dad leaned over the teacher's desk as she was pointing out something in her grade book. I could see by the expression on his face that he wasn't too happy and my mom kept glancing over her shoulder at me, with her eyes shifting back and forth. Thank God I didn't hear any huffing sounds from her. Then, without warning, the teacher, Dad, and Mom, made their way down the aisle to my desk. I looked up, "Hi, Mom. Hi, Dad," I nervously muttered. Chuckles started to circulate around the back of the classroom.

"Let us see your notebook," my teacher demanded, "so your parents can see your work." I slid my book over to the side of my desk and then my teacher opened it up. First she looked at the social studies section, then science, English and then my secret-coded section. The teacher turned the tab and there appeared my 'doodles', page after page. The '*Nina*', '*Pinta*' and '*Santa Maria'* in their glory o'er the ocean blue, then my best drawing of a British man o'war, a brig, a frigate... on and on, page after page. My dad's face was now expressionless. He remained very coherent, though it seemed he might be gasping for air. Mom had started her huffing and my teacher was looking at me as if to say, "I got ya." My dad's upper lip started to disappear. I knew that gesture and what it meant. This will be the end of me. I was so mad at my teacher, cringing at the idea that she had called them for a classroom conference. Saved by the bell, we all had to go to our next class, no time to dilly-dally; we only had three minutes to pass. "Good-bye, Mom; good-bye, Dad; see you when I get home." I'm sure I will!

Well Dad got home from work, we were all ready for dinner and it turned out to be a quiet meal. When we finished and were excused, each of us had chores to do: washing the dishes, rinsing, drying, sweeping the floor, and my favorite, washing the pots and pans. When you're in charge of the pots and pans, no one bothers you. Actually no one will stay near you for fear you'll ask them to help out. The only interference that job had was the pots were subject to Mom's inspection. She would commonly instruct for this task "More elbow grease," she'd insist.

After I was finished, Mom asked me to go sit down next to Dad while he was having his tea. Okay, I thought, it's a delayed reaction. I went into the dining room and sat down next to him. Mom stayed in the kitchen. Not a good sign, I thought. Between sips, Dad told me how disappointed he was with me and that I really needed to buckle down and put my nose to the grindstone. I wondered about what that meant. We didn't have a grindstone and why would I want to put my nose on one. To ensure my homework was getting done, he ordered me to do my work at the dining room table. He would check it when he came home from work every evening.

"I don't want to have to go up to school again and find out you're not doing your work. Now go up to your room and clean out your notebook." I did what he asked, took out my 'doodle' section and put it in my desk for safekeeping. When I was finished I had Dad check it out, "Very good, Son, that's the way I want to see it from now on."

Columbus Day came and went as did Armistice Day and Thanksgiving. I was hoping I could take a bus on the weekend up to see Gus and Skipper but Mom said I had to get permission from Dad. I asked him and, with Mom's support, he said, "Yes."

I knew just how to get there. Take the Greyhound Bus that said 'Bennington' and ask the driver to let me off at Ed Jones Road. I called to find out what time the bus was leaving, took the BMT to 34th Street, walked over to the terminal booth purchased my ticket and waited on the platform. Eventually a big Greyhound bus pulled alongside the platform and opened its doors. I was the first to get on. The driver checked my ticket at which time I asked, "Could you please let me off at Ed Jones Road?" The driver thought a minute, then told me, "Have a seat right here behind me. You'll have to let me know when we get near there." The bus ride was a lot of fun and the driver was friendly.

As we rode further from the city, more snow began to appear on the sides of the road. I thought to myself how surprised Gus and Skipper will be to see me. All kinds of ideas ran through my mind as the bus wound round and over the curvy New England highways. Every once in awhile the driver would lean back and talk with me. I told him I had friends up in the Berkshires and was on my way to visit them. He was a nice guy, cordial and interested in his passengers. He even carried elderly passengers' bags for them when they got on and also when they reached their stop.

We stopped at a rest area in Danbury but I stayed on and just waited for everybody to get back on the bus. I was anxious to keep on going. The snow was getting deep, the trees all bare and stark gray standing amidst the snow-covered ground. The bus was scheduled to arrive at the General Store at 1:25

p.m. Ed Jones Road was only five minutes past the store on the left. "We're almost there," I said as I leaned forward looking over his shoulder. "Just a little ways up," pointing with my finger. The road sign came into view. "There it is," I said as the driver stepped on the brakes. A rush of air resounded from the air brakes and the bus pulled over onto the shoulder and the doors swung open. I grabbed my bag and said "Thanks! See you day after tomorrow."

There was Gus sitting on the milk platform, He greeted me with one of his firm handshakes. I knew better this time not to grab his hand first and then he followed with a big hug that almost took my breath away. "Good to see ya', mate," he said with a big grin on his face. "Same here," I piped up.

We marched up the hill and onto the flats and then passed by our summer house. It looked so deserted now. A few more months and it would be humming again with activity. A short ways up the road, there was Skipper greeting me at the door with a sandwich and a cup of hot tea for lunch.

"How did you know I was coming?" I asked Skipper. I wanted this to be a surprise visit. She looked at me and just smiled. I guess Mom must have called her to let them know that I was on my way.

Gus sat with me at the table and discussed the weather as I 'et' my sandwich. I was happy to be there. Everything looked so different with the leaves off the trees. The fields all looked bigger and were now dressed in a blanket of snow.

The horses were walking around the barnyard not looking as sleek as I last saw them. They had all grown long winter coats. They looked as surprised as I was. How different it is up here in the winter with the cold and wind; how different from the pleasant warmth, green trees and fields of summer grass.

Skipper had the cook stove going and it was warm and toasty. Finio, their dog was curled up alongside the fire. Now and then the stove would burp smoke and fill the kitchen with the fragrant aroma of cherry and birch wood.

Gus put my knapsack into the parlor, the big room right off the kitchen. It had its own wood stove set in front of an ancient fireplace bedecked with old paneling. "This is where you'll bed down, mate," Gus said, pointing to a puffy old couch against the wall. "Now, come on, we'll go out to the barn and see our buddies."

When Gus slid open the door, a rush of warm air pressed against my face. There was Bessie in her stanchion munching on hay. Bos'n, the duck, the chickens and, of course, Lymey, the goat, were all busy eating their vittles. The horses came up to the door and wanted to come in and see what was going on. Gus put them back in their stalls and put a flake of hay in their mangers. It was so warm in the barn; the animals were all safe and cozy. The smells and sounds of life around a barn are a sensory experience that I have always cherished.

Gus slid the door shut, stopped by the well, pumped a bucket of water and back to the house we went to 'chew the fat' as he would say. I didn't tell Gus about my parent-teacher conference. He didn't think I paid attention in school anyway.

Gus went into the parlor and stoked the wood stove. It was getting late and time to hit the sack.

"Here ya go, mate," he pointed to the couch. "See ya in the morning."

I buried myself under the old wool blankets set on the couch. As the door to the kitchen closed, the parlor became pitch black except for a beam of light from the fire reflecting on the ceiling. I thought I would never fall asleep. The tick-tock of the mantle clock seemed to get louder and louder as the night wore on. The old beams creaked from the winds as they buffeted against the house. Before I knew it, morning came and I could hear Skipper starting up the cook stove in the kitchen. 'Va-voom', a little cup of kerosene will do ya. I peeked out through the top of my blankets and was blinded by the sun glaring through the east windows. I took a deep breath, felt the coldness in my nostrils, and as I exhaled, a cloud of steam ushered out and hung around my head. The stove had gone out—obviously a long time ago; the windowpanes were all covered with frost. I could smell the coffee drifting in through the crack in the door. I stepped into my clothes, which were stiff as a board, hopped around looking for my shoes and quickly made my way into the kitchen. Rubbing my hands I stood by the cook stove to get warm. Skipper placed a large skillet on the stove, put in some butter, which immediately sizzled. Gus came out of his room and also stood by the stove trying to catch as much heat as he could. Coffee was done and we sat at the table embracing the hot cup in our hands. "What's for breakfast, Skipper?" Gus piped up. "Swedish pancakes," Skipper replied.

"Aye, aye, Sir!" Gus saluted. Along with the pancakes now underway was a skillet of sizzling bacon. The kitchen was full of great smells. The Swedish pancake was made with scrambled eggs cooked like a gentle omelet, turned over, cooked on the other side until just slightly brown then slid on a plate, covered with strawberry jam, rolled up like a jelly roll and then cut into slices. A few slices of bacon, a cup of hot strong 'java' and it was like I was in seventh heaven.

"Let's go, mate, time to do the chores." Putting our coats and boots on, out we went. The sun was breaking high over the hills and few, if any, clouds were in the sky. A beautiful day was coming up. Gus slid open the barn door and as usual we got a boisterous greeting. Out came the grain, the mash and the hay. Buckets of water were fetched from the well. "Hey, mate, bring that wheelbarrow and grab the manure fork." I wheeled the barrow in and I figured this was going to be my chore, mucking the stalls. Up and down the fork went loading the manure onto the barrow. Each full load I pushed out to the manure pile. It was so warm that I had to take my coat off. Gus was getting ready to milk Bessie. I stood clear this time. Gus finished his milking without event and back to the house we went. "Don't take off your coats, lads. I need some firewood," Skipper demanded. We went into the woodshed which was right off the kitchen. There was no wood in it just tools and cans of nuts and bolts, a sawhorse, a grinding wheel, a bucksaw and a two bit axe sticking out of a stump. Gus grabbed the axe, and out the woodshed door he went, with me following. "Where are we going?" I asked. "Never you mind, just follow me!" Gus, with me following closely behind, went across the meadow with axe in hand. Gus kept looking and shifting his glance across the wood line.

"Ah, ha, this way mate, follow me." At the edge of the field, Gus found a straggly-looking birch tree and, taking a stance, raised the axe over his shoulder and began chopping. It didn't take very long to chop it down as it was only three or four inches in diameter. He trimmed the few branches and we were ready to head back. "Here, mate, you take the axe and be careful; it's sharp." He lifted up the tree, balanced it on his shoulder and back to the house we went. "Grab that sawhorse, mate, and set it here by the door!" Gus set the end of the tree on the sawhorse, grabbed the bucksaw and began sawing it stove size.

After sawing up an arm full of wood, we carried it into the kitchen, leaving the rest of the tree on the sawhorse, the butt end sticking out of the door. "Here ya go, Skipper," Gus said, stacking the wood by the stove. We sat down at the table and were trying to figure out what to do for the rest of the day when Skipper piped up, "Don't bother to sit down. I need more wood than this. How many times do I have to tell you, Gus? The fire has to be kept going. It's not just for your chow, you know; we have to keep this place warm!"

Gus looked over at me with a grim expression on his face, and rose up from his chair. I followed his lead out to the barn. Gus grabbed the harness and set it on Boy. "Go get the axe, mate, and meet by the ash tree." I got the axe and along came Gus leading Boy in hand. Off we went again into the field toward the edge of the woods.

"There's a couple of nice birch and ash trees out here," Gus muttered as he scanned the woods. "They burn good, ya know, even when they're green." Chop, chop, Gus swung his axe.

One by one the trees fell until he had four, then he trimmed them all ready to go.

I brought Boy up, Gus wrapped a chain around the logs and Boy walked forward until the slack was out of the chain. Back toward the house we went. Back in the barn went Boy after we unloaded the logs by the woodshed door.

One at a time Gus sawed them up, taking most of the morning and afternoon while I stacked the wood neatly in the woodshed. All told Gus had cut approximately one half a cord of wood.

By late afternoon time, the wood was all stacked and ready to go.

Now it was time for the evening chores. As the sun began to set I could feel the coolness rush in, Gus said it was another cold night coming up.

The animals were gathering at the barn door as if they had a clock in their head and also knew it was dinner time.

The horses were put in their straight stalls and tied. Hay was thrown into their mangers and Bessie the cow walked into her stanchion, waiting for Gus to close it. She had her portion of grain and hay as well.

Porky managed to be in the barn but had to be escorted out because he thought if he mingled in this barn he would get fed and then could scoot back to his pig sty to get fed again.

On the way back Gus noticed some critters hanging on for a free ride and I'm sure a free dinner. Gus went to work throwing them overboard. This is not exactly what Porky expected. 'Where's the chow, Jack' he oinked as he patiently waited.

Then over to the duck house with a pause for a brief chewing out by Gus for the racket the they made that could be heard all over the hilltop.

Ducks have no patience you know, nevertheless everybody was finally tucked in.

Skipper rang the bell; dinner was ready and waiting. On the table were two big plates of beans with a couple of pieces of bread. I was so hungry I almost began before Gus said his usual grace, "God is good; God is great. Thank you Lord for these beans on me plate."

We both made quick work of the beans and I remembered to ship my oars when I was finished this time.

Gus was very fond of his beans, a regular on his menu. I liked beans too, but they always gave me gas. That's why I never ate them at school when served with hot lunch. Many of my classmates seemed to avoid the beans, too. The afternoon lessons in class would always be interrupted by someone murmuring, "Who did that?" with many disclaimers following.

Gus got up, went into his room and came out with something in his hand. Whatever it was, he sat on it. We were talking, just discussing things when all of a sudden, I heard this loud 'blat, blat!' I was surprised and embarrassed as it sounded like Gus just passed gas. He leaned sideways and again, another 'blat, blat.' Skipper acted annoyed and said, "Gus, that's not a nice thing to be doing in front of a young boy!" I glanced over at Skipper trying to figure out what was going on when Gus pulled out a contraption from beneath his seat. It looked like a slingshot made out of a clothes hanger; a rubber band with a big flat metal washer in the middle. He held it up and wound the washer around and around on the rubber band, lifted up in his seat and sat on it again. After a few seconds went by, he leaned in his chair and 'blat, blat.' Off it went again. I almost died holding back my laughter. It wasn't very polite what he was doing but you just had to be there. His piercing blue eyes and that smirk on his face gave way to the impish side of him. Skipper was fuming and just glared at him. Gus stopped laughing and straightened up. "I guess that's enough of that," he said.

Gus got up went back into his room and this time he brought a 'little blue book'. He set it on the table and turned up the wick on the kerosene lamp. Gus was wearing a sleeveless tee shirt and the brightened light brought to life the tattoos on his arms and chest. I took this opportunity to ask "Gus, where did you get all those tattoos?" I pointed first to the one on his right arm of a dagger stuck half in and a drop of blood running down his arm.. "Oh, that one I got in

Cardiff." Taking up the lead, he now asked me, "What do you think of this one?" Pulling down his collar he exposed a big sailing ship and the British coat of arms with flags crossed. Then turning his left arm so I could see a lady holding a spear and cross flags, he said, "She's the Queen of the Sea and these are the ensigns of the British Empire." On his hand was an anchor just barely visible and pulling his pant leg, he pointed out three round brown scars. 'What are those?' I asked curiously. They did not look like tattoos. "These I got from an octopus in the South Pacific. Those marks are what he left me as souvenirs. Well, it's been a long day and tomorrow t'is another cold day a coming up."

I was so happy to have spent the weekend up here and to see what the farm was like in the winter. It was powerful. When the wind blew it howled, when it snowed it was deep. Nothing like the city where most of the time there were just flurries and the wind was no more than a breeze.

The kerosene light began to flicker; it was getting that time to go to bed. In the winter up here, it gets dark so early that when 5:00 o'clock rolls around it feels like two o'clock in the morning. Gus stoked the parlor stove, closed the door and I settled in for another chilly evening. It was hard to sleep because it was much colder than what I was used to. The couch I was trying to sleep on was stuffed with horse hair and was itchy and hard. Still it didn't take long to fall asleep.

By morning there was thick frost on the windows and no sun, just lots of clouds. It looked awfully bleak and dismal. Skipper had the fire going and I stepped over to the stove to get warmed up. Gus was putting on his boots and getting ready to do the chores. Skipper was cooking but Gus was in a rush to get down to the barn. So, I had doubles; his eggs along with my own. I had to be off early today to catch the bus back to the city. After breakfast, I said my good-byes and thanked Gus and Skipper for the wonderful weekend visit. I told them I'd try to get up again. I reminded Gus that he didn't go through that little blue book of his. He promised next time for sure. I grabbed my knapsack and headed on down the road to the milk platform and waited for the bus. When I heard it rounding the corner, I stood up on my toes and flagged him down.

"Hi. How ya doing?" the driver greeted me, "Climb aboard." I got to know him over the next few years as this was his regular route and we became good friends. He said "I saved you a seat behind me." as he pointed over his shoulder. The doors closed and off we went back to the Big Apple. The closer we got to the City, the heavier the traffic and we had to limit our discussions. We pulled into the 34th Street Terminal. I thanked the driver and explained I'd probably see him again in a few months. The BMT to Brooklyn was crowded as the rush hour had begun. It was a far cry away from the quiet and calm of Gus and Skipper's place.

Monday morning came and I was all set for school. I did my best to pay attention and do well since I had plans to go up to the country again soon. It wasn't long before I was back to doodling but I also took good notes in class and even copied work off the blackboard.

As the winter progressed, I wasn't able to get back to see Gus and Skipper. We always had a big Christmas and everybody was expected to be home for the holidays. Spring came and Easter passed as the end of the school year approached. We kids started to get antsy in anticipation of going up to the country again for the summer. Mom had begun to put things aside. We knew once we got there, we'd have lots of work to do. Mom was sure the mice had moved in and raised havoc with the house. The inside of the drawers, dishes and silverware would all have to be cleaned and washed down with Clorox.

Finally, the day came, everything was loaded in the car. We sat everywhere in the back—on the seats, on bags of clothing, even on the back window deck. Only one of us would be lucky to get to sit in the front seat and be first to spot the Ed Jones Road sign.

It was a long trip as usual. We made up games, identified license plates, waved to people to see how many would wave back and even counted red cars. Bonus points were available for the one who spotted a police car. Time passed, some of us slept and then were awakened by Henry yelling, "We're here!" He was the first to spot the road and got the fifty bonus points.

Dad down-shifted as the car rocked onto the old dirt road with all its bumps. Up the hill we went, across the flats and straight away to our summer country home. There it was standing in the midst of green grass. We tip-toed up to the front door, making sure not to disturb any creatures. The house was musty and damp so we immediately opened all the windows. Back and forth to the car we began: carrying groceries, clothes, and all the other things that Mom had packed to make her job easier for the summer.

Supper that evening was simple. Mom made sandwiches, knowing she couldn't cook until the house was scrubbed. We all sat around looking everything over. We checked to see if our toys were in the same place and planned for tomorrow's activities. We made our beds, laid out our clothes and settled in for our first overnight of the summer.

I woke early to the birds chirping in the light mist that hung over the hill. There was much to do. The grass had to be cut, the fire pit needed repair, and I had to figure out when I could go up to tell Gus I had arrived. I hadn't seen him since before Christmas and was anxious to go up the road. I was sure all these chores had to be done before Mom would let me leave.

To my surprise, Skipper walked down the road and spent some time talking with Mom over coffee. I slipped away from my chores with Dad on the pretense of needing something to do in the house and walked through the dining room. "Hi Skipper," I said and she reached over and gave me a hug. "Where's Gus today?" I asked. "Oh, he's at work today and will be home at five o'clock. Why don't you come up after supper? I'm sure he'll be happy to see you." That invitation was just what I needed. Mom said nothing, so right after dinner, I raced around getting my chores done, then I went up the road.

"Hi, mate." Gus greeted me at the door with one of his crushing handshakes and breathtaking hugs. I sat down and settled in my chair. "Come on, mate, get 'em up." his arm posed hand open. I grabbed a hold and hung on, his fingers wrapped around my palm and with a squeeze of his thumb and three and one-half fingers, down I went, my hand slamming on the table. He grinned and again reminded me he was the bos'n.

We talked about the events of the winter. The two Northeasters they had in February resulted in thirty inches of snow, drifts as high as six feet and temperatures below zero. Summer was here now and there was a lot to do. We had to mend fences, get the equipment ship shape for the haying and, hopefully, Gus would show me his little blue book. "Maybe, mate, you could give me a hand one of these days with the barn roof." I've got to set and tar the nails." "Great," I thought, "sounds like a good project, up aloft on the roof. "Sure, Gus, you can count on me! By the way, Gus, do you think you could take me fishing one of these days out in the lake?"

"Sure, mate, we'll rent a rowboat and I'll show ya how to row. Let's see what tomorrow looks like." I hung around awhile as the peepers began to broadcast over the hill. I knew I had better get home because tomorrow Gus and I may go to 'sea'.

Arriving back home, I searched out Mom to ask if I could go fishing with Gus the next day. "Okay," Mom said, "but when you get home I have things for you to do that just have to get done." I kissed Mom goodnight, anxiously waiting for morning to arrive to 'ship out'.

I loved to go fishing and spent many an early morning in the city out in Sheepshead Bay with one of our neighbors. He had an outboard motor and made the best 'chum' in Brooklyn. I'd be up and ready at 4:00 a.m. in the morning. We'd go out to the marina, rent a boat and load in our tackle and 'chum'. Many times the water would be rough and there would be good size swells. We would turn the motor off, drop in our lines and drift along with the current. Then we would start the motor, go back and drift again. I used to see big freighters heading through the channel. We were like ants next to them. Their wakes would rock our boat back and forth until they were long past.

Early morning came and I was up. I had brought my fishing gear and tackle box from the city even though it was salt water gear. I had some pretty flashy spinners that I felt should work well in fresh water. We got a ride down to the marina from one of Skipper's friends. Gus checked out the row boats, specifically looking for one that didn't leak. That was his department. He found one that would do and we loaded the gear on the boat. Gus set the oars in the locks, pushed off from the dock with the end of an oar, and we were underway. I sat up front watching the water ripple past the bow as Gus rowed out to the middle. The water was like a mirror. Once far enough out Gus pulled in the oars. I cast out my line and reeled in and cast out again—nothing, not even a nibble. Gus took out a line wrapped around a stick from his pocket, put a worm he had brought from home onto a hook, attached a cork from an old wine bottle for a bobber and dropped it in. It drifted a ways and every once in awhile I'd see little ripples around the cork.

Gus set the oars and began rowing again. After a little ways, he shipped the oars, got up and said, "Here, mate, now it's your turn." "Wow," I thought "Gus is going to teach me how to row." We exchanged seats and I grabbed hold of the oars. They fell out of the oar locks and I struggled to get them back in. I started to lower the oars and had a hard time just keeping them level. One oar was up, the other deep in the water. "Now, mate, push down a little so they're both riding on the top of the water," Gus suggested. I improved after a few attempts and was able to get the oars out of the water. With each pull the row boat moved ahead. The uneven pulls caused the boat to shift erratically in the water, but at least the boat was moving. "No, mate, easy. Move 'em easy" Gus echoed in my ear. Gus stiffened his hold on the side of the boat, trying to keep his balance. "No, mate!" Gus exclaimed getting up out of his seat, "Let me back in there." He now sounded even more frustrated. I slid over and exchanged seats forgetting to ship the oars. One oar slid into the water. "Cheese and rice," Gus muttered, stretched to grab it and just missed it as the oar drifted away. Gus grabbed the other oar and rowed the boat from the stern like a gondola toward the floating oar. As he reached over and grabbed it, he looked back at me shaking his head.

Just then a power boat came by pulling a water skier. It was turned our way, with the water skier skimming the water sideways, throwing a large wake our way. Gus was still standing and setting the oar in the lock when the wake almost tossed him overboard. I could hear him now, "Man overboard!" Now, Gus was really mad; shaking his fist and with his other hand waving the oar in the air and letting go of the longest train of expletives, similar to what my dad says when someone cuts him off on the road. Again, words I knew I should neither repeat nor discuss any further.

The boat swayed and rocked back and forth and I held on to the seat. Gus was still standing, rocking with the sway of the boat like a bird in a tree during a hurricane. The water gradually settled and Gus proceeded to get back to the business of teaching me how to row. "Now, watch me, mate, carefully." He explained I should pull the oars toward my waist pressing down a ways, just to keep the oars above water. Then I should push the oars forward, keeping the handles down like so." The oars looked like wings on a bird as he rowed evenly and smoothly. "Now lift up easy and make sure the blades are at an angle. You can do that by turning your wrists together. Give them a little twist as the oars enter the water and once in the water, not too deep mind ya, pull back evenly toward your chest and twist your wrists down like this."

Gus further explained, "When the oar handles touch your chest, twist the handles downward, the blades of the oars should be horizontal with the water. Then push away from your chest and down toward your waist. Don't forget when you are not rowing, always pull the oars in and drape them over your lap or ship your oars when your leave your seat."

Gus rowed across the lake, each pull of the oars smoothly pulling the row boat forward. The oars glided back and forth over the water in a steady rhythm, the bow of the boat cut through the still waters. “Watch what I do; this is called feathering your oars,” Gus pointed out as the oars skimmed across the water. We rowed around the lake like we were on an ice sled. “Now, it’s your turn.” Switching seats again, I took the oars and practiced rowing for the rest of the afternoon until it was time to head in. It’s a good thing Gus remembered which shore the marina was on. We had gone around so many times I had lost my bearing.

We hitched a ride back to Ed Jones Road and walked up the hill and back home, “Thanks, Gus, I’ll try to get up to see you later this evening.” I said as we parted.

I stretched out on my bed, my arms numb from all the rowing and fell asleep, even missing supper. Mom didn’t wake me because I was sleeping so soundly and I didn’t awake until morning. I wrote Mom a note: “Dear Mom, I went up to Gus’s. Be home for lunch for sure. Love you.”

I saw Gus sitting on the well head with a bucket of soapy water at his feet and in one hand part of the harness which he was scrubbing with a sponge. “Grab a seat, mate, and give me a hand with this.” One of the many jobs in preparation for haying, I thought to myself. “Take this can of oil and start rubbing it into the leather. Use your fingers and rub it good.” Gus handed me the oil and I went to work. We sat there for the better part of the morning cleaning and oiling the harnesses and bridles. Skipper stuck her head out the door. “Lunch is ready.” I excused myself because I promised Mom I’d be home for lunch. “What are you doing this afternoon?” I asked Gus. “Well, mate, I think maybe I’ll take on the barn roof; come up if you can.”

I finished my lunch and scooted back up the road. Gus had the ladder out leaning against the barn. The front of the barn wasn't steep but where it connected to the hay barn, it had a sharp pitch. Gus grabbed his bucket of pitch, a can of nails and stuck his hammer in his belt. Up the ladder he climbed, "Follow me, mate." I followed him up and perched myself on the low roof. "Mate, you hang on to the pitch bucket while I start resetting the loose nails."

The barn roof was made of tin corrugated sheets and I could see many nails had popped up. Leaning on one leg, I shimmied across the roof while Gus hammered down the nails, took the stick from the pitch bucket and dabbed each nail head.

We worked one end to the other. When Gus finished the lower roof, he grabbed a large rope that he previously had thrown across the steep roof I hoped it was anchored on the other side. With one hand, he pulled himself up. In the other hand he had the bucket and nails. He perched himself on the peak and started walking across the ridge. Boy, I thought, did he have good balance. With rope in hand, he straddled down the roof and finished the job.

The sun was getting pretty hot and the tin roof even hotter. I was no longer able to sit on it. "Gus, I have to get down," I said. Gus hollered back "Okay, mate, I'm almost done, be right down."

I climbed down the ladder and found a shady spot under the ash tree. Gus came down covered with rivulets of sweat. After we put his tools away, we leaned back against the tree.

Skipper, as luck would have it, brought out a pitcher of ice cold lemonade. We sat relaxing on the grass.

Bos'n, the duck, paid us a visit. A few quacks, a little surprise and off he went about his business on his tour around the barnyard. Everything was his business and if he noticed something not routine, everyone around knew it.

Later, "Hi Mom," I greeted her after a hard day's work. "What's the black stuff on the back of your pants," she asked. "Uh oh! It must be from the roof. I was helping Gus fix the barn roof," I admitted. Here came Mom's look, shifting her eyes and now the huff, "You were what! Go to your room and take those pants off right this minute and bring them down to me. Those are your good pants."

Mom worked on my pants. I didn't realize how hard it would be to get the tar off. I went to her garden and tried to make myself useful. Evening came and Mom was still upset so I stayed home and worked a few problems Dad had circled in my math book.

Gus's stories occupied my mind over and over again. Images of sailing ships danced around in my head. I daydreamed and imagined what it would be like aboard a fully rigged vessel.

Being around Gus and helping with his chores, I learned many things about carpentry, horsemanship, care of animals, farming equipment and general repairs on the farm. Gus had limited resources; he had to make do with what he had on hand. He had a huge collection of nuts and bolts, jars of bent nails and pieces of wood planks in every width and length, all neatly arranged in his woodshed.

Every job he did around the farm was 'make do.' His resourcefulness and inventiveness kept him on top of what needed to be done. I was learning quickly all the ways of his clever existence. Every tool, every nail was kept in good condition. Gus would straighten old nails and reuse them, make gaskets and bushings from old tire casings, bend old iron to adapt to needed repairs on his farm equipment. His perseverance in creating alternatives was what I later became familiar with in the Army. This was called 'field expediency' and Gus was the master at it.

Gus taught me how to use wrenches, break bolts, and use leverage to lift huge stones just like the Egyptians. He taught me how to dig a trench around a large rock, lift the rock with a chain hitched to his horse and inch it out by backfilling the hole so that, before you knew, the stone sat on top of the ground ready to drag away.

Another example of his perseverance and ingenuity was shown when my mom wanted to put in a running water system using a Ram pump in a spring located about six hundred feet behind our summer house. A trench needed to be dug below the frost line. In this area the frost line was at four feet.

This required digging a trench by hand because at that time there were no bucket loaders to do the work. Gus organized a crew of four men with hand shovels to dig a trench three feet wide, four feet deep and six hundred feet long to the spring. This required moving many large rocks. Together with his team of horses and his tackle blocks and chains Gus managed to successfully dig out a straight run to the spring. The Ram pump worked on the principle of gravity. Set below the spring, the water would travel down a chute. After gathering sufficient force the water would enter the Ram pump and be propelled up the water line to a tank by the house. This was quite a feat. We were the first on the hill to have running water, no longer having to pull water up out of a shallow water well. We had pure spring water. After a few years, the spring collapsed on itself, forcing us to install an artesian well run by a gasoline engine on a Meyers pump. Not having any electricity, we would have to start the pump every time we needed water. Still, it was easier than carrying it.

I later found that Gus would work about two weeks a year down at a paper mill long enough to make the money to pay the annual real estate taxes. He also worked odd jobs, usually just for neighbors. He didn't drive and I never saw him operate any machinery other than horse drawn. Lifting and pulling were done with horses and blocks and tackle. Field work was done with his team of horses.

4

The Little Blue Book

Finally, under the glow of the old kerosene light, there was the 'little blue book' sitting on the table. I leaned forward in my chair and studied its cover. Hopefully tonight Gus would reveal its contents to me. The cover was blue cloth and in the center a little box with his name: "Ture Alfvegren" faded and hard to make out. The cover was like a canvas wrapper that went around the whole book. Gus opened the cover and on the inside was a crown and a circle which read "British Board of Trade" above a serial number and words reading "Merchant Shipping Act 1904."

Gus started to read, "The Master shall sign and give the Seaman a certificate of discharge. The certificate must be in a form approved by the Board of Trade, specifying the period of service and the time and place of discharge and if the Master fails so do shall for each offense be liable to a fine not exceeding ten pounds."

Gus's finger followed along with each sentence sometimes hesitating as he pronounced a word to be sure it was correct. Gus continued to explain the importance of this book. "It contains the names of every ship I served on, its tonnage, the journey, the captain's name and the ports that I sailed to. Inside the little blue book were two other smaller books. Gus took out the one which was entitled "SJOFARTSBOK." Gus said, "This is my Swedish book which I got when I first became a full fledged sailor. See here," Gus pointed, "Jungman, a sailor." Turning the page he read off the name of the first ship, *'Emma'*. "She was a schooner. Here, mate, the '*Fredhen*', she was also a schooner. We sailed cargo along the coast to Stockholm and then to ports in the Baltic, the Gulf of Bothnia and the Gulf of Finland." Gus turned the pages in his atlas to a map of the Baltic, pointed out the many ports and then pointed down at the island of Gotland.

"Here mate, is Wisby, that's where I come from, remember? I sailed all these parts as a young boy, to Finland and even Russia. Here is Hamburg in Germany, I made many voyages there also." Gus stated proudly.

"Let me tell you something about Gotland. It was an independent country until 1679 when it became part of Sweden. My people lived here for five thousand years. It was also a Viking trading station, an important center for commerce in those days. My family descended from the Vikings." Gus's eyes lit up, a proud grin spread across his face.

Banner
of
Gotland

Gus took a deep breath, puffed up his chest and continued, "There is a big stone wall that surrounds Wisby, built to protect the people from sea raiders. One of the most important raids was when Danish King Valdemar landed with his army on the west coast of Gotland. The peasants armed themselves and eventually drove him out."

Back to the log book, "Here, mate, I signed on the frigate *'Amsterdam'* in Stockholm. I was promoted to mate. We sailed to Norway then up the coast to the northern ports, so many different ports up there. She was a bigger ship and I made a lot of voyages on her. Norway was beautiful with all its inlets and fjords." Checking the dates in his log, he said "I spent about a year aboard her," he read aloud. My next voyage was on the bark, '*M.Roosvall'* as a mate, too. I was still pretty young then, about seventeen or so. We sailed to ports north and east all through the Baltics for about six months."

"See here, I did a tour on the 'Reserve' and the '*Burgermeister Petersen'*. Those were exciting days; so many places to see and so much to learn. After a voyage to England, I registered with the Board of Trade which enabled me to sign on to any British merchant ship. That's the little blue book, "Official Maritime Register."

"I was now able to sign on to any Scandinavian or British merchant vessel. I got my experience on sail ships and was now ready for steam ships. I started 'steam' as an able-bodied seaman. Look here, mate, these are the different stamps for immigration. When at port for an extended period of time, I had to go and report to the police or immigration bureau."

In front of the blue book were also consulate stamps and identifying information. It even listed Gus's tattoos. "All sailors had to carry this book with them at all times. It is my record of work and reference for how well I served on each ship." Turning the page, Gus pointed to the first entry. "The name of this ship was the '*Rotterdam*', three thousand one hundred and thirty tons, a good size vessel. I signed on in Rotterdam, Holland, with the rating of able-body seaman. We sailed from Holland to Calcutta, India, then on to the United States and back again to Rotterdam. I served aboard her for three and a half years. She was a fit ship. Anyway, we left port and headed into the North Sea: the weather was stormy with gale winds and very rough seas. I remember how big the swells were in this part of the ocean. This was my first experience out in the Atlantic. The bow would sink down into the swell, then lurch back up to the crest. Sometimes the waves would break over the sides spewing water across the deck. We had to be very careful on deck that a wave didn't break over the top and take you down. That was my first day on a steamer, a little different then riding it out on a sail ship. I just wasn't used to those steam vessels yet. After three hours or so, the winds subsided and the seas turned calm as we headed southwest into the open waters of the Atlantic. This is where I got my steamer sea legs."

"We steered down along the French coast, past Spain, then on to West Africa, heading to Monrovia in Guinea. There we dropped anchor in the harbor. We went ashore for fresh stores and laid over for two days while the engineer made some repairs. We set sail at the tide and crossed the Equator. It was my first crossing so the old salts made me wear a pirate's outfit. There was a little celebration like an 'initiation' for those who never crossed the Equator."

"We had to stand on the gangway extended out over the side, and guess what? We had to jump in. Yes, in the middle of the ocean. Sure, the old salts plucked us out but still, treading water in the middle of nowhere…. what about the sharks? Believe me, we kept our eyes peeled. Some of the crew had rifles which the captain passed out from his armory just in case. Nevertheless, it was scary."

"I was now a veteran, I had crossed the Equator. This was the first of countless journeys I would make passing across." Gus got up and brought from his room a small picture album and turned to a picture, "What do ya think? Do I look like a pirate?"

"Great picture," I said laughingly looking at the expression on Gus's face in the picture.

"Well anyway, we continued across the Equator, followed the coast of Africa toward the Cape of Good Hope, which is at the southern most tip of Africa. The seas got very rough. We braced ourselves as we sailed around Cape

Colony, where the south Atlantic and Indian Oceans meet. The seas were confused and we tossed and turned, plowing through the swells and high waves, twenty or so feet high. We passed into Delagoa Bay and steered east of Madagascar passing close to Mauritius. It used to be called the Isle of France, now it's a British colony."

"Are you following me, mate?" I nodded and Gus continued. "We then steered north-northeast toward Calcutta, across the Indian Ocean." With his finger, Gus traced the route the '*Rotterdam*' took. I guess he expected me to remember all these places. Gus added. "The sun never sets over the British Empire. No matter where in the world you went you'd almost be certain to find a Union Jack flapping in the breeze."

"We finally sighted Ceylon off the southern tip of India and then steered north toward Calcutta. At the mouth of the Hooghly River, we dropped anchor. Calcutta was a very large city; over two million people back then and maybe doubled today. Here cargo was unloaded onto smaller vessels and we took new cargo, refueled and replenished our stores. We left with the tide to our next destination."

"South we steered, then east through the Indian Ocean. We ran into a typhoon just before we reached the Sunda Islands. A furious storm, the winds were fierce and the waves monstrous. It tossed us for hours and the captain tried to sail around it. We had all we could do even to make way. As we passed through the islands, the seas eased up and the winds died down. We stopped in British Borneo, part of the Lesser Sunda Islands to repair the damage to the rudder caused by the storm. We even sprung a few leaks, so we laid over while welders made the repairs. We set sail on the fourth day and steered northeast toward the Pacific, then across the Equator again. Ha, ha and then on our way to San Francisco in the United States."

"Just before we entered San Francisco Bay, we waited for harbor pilot to bring us in and escort us to our pier. San Francisco is a beautiful place. We spent several hours ashore just trying to get our land legs back again after our long voyage. We laid over for several days while the cargo was unloaded and a new cargo taken on. The return trip to Holland would take us along the coast of South America. We dropped anchor in Valparaiso for fresh stores and a short layover while some of the cargo was taken off. The engineer had some problems with the boiler and as soon as it was repaired, we continued south to Cape Horn. We rounded the horn amidst high seas with waves topping forty feet, commonly referred to as the rollin forties."

"It was like entering the gates of hell. It was a fight for our lives to get out. After several hours of hard sailing, we finally passed Tierra del Fuego, south of the Straits of Magellan and east of the Falkland Islands. We hugged the eastern coast heading north and we stopped briefly in Rio for fresh supplies and fuel. On our way again in the Atlantic, we headed northeasterly toward London and then on to Rotterdam. This was my first all oceans sea voyage." Gus retraced the route again on the map and pointed out the places he had been while with the *'Rotterdam'*. "I signed on July 6, 1904 and was discharged in London on January 26, 1908. The captain gave me a very good rating which helped me when I signed on to my next vessel."

"I lay over for a month just relaxing and seeing the sights. I eventually got bored and was aching to be back to sea so I signed aboard the *S.S 'Farringford'*, a two thousand and seventy-eight ton vessel. Our destination was

Port Said. We sailed south to the Mediterranean, then east to Egypt. Port Said is a good size port but you have to be careful when you go ashore. Anyway, we did our business and headed back west. Along the way we made a brief stop in Lisbon for a minor repair where I spotted an old friend ready to get underway, a three-masted schooner loaded high over the gunnels with logs. It was now quite a while since I had seen one of those good old sail ships."

"This was a short voyage, let's see, it says I left in February and returned in April. That's about three months. A week later I signed on the *S.S. 'Competitor'* as an able-body seaman and departed on May 7, 1908."

"The *'Competitor'* was a fine ship, sound and seaworthy. We left with the tide but after five days out or so, we ran into a storm and immediately starting experiencing problems with the boilers. Finally the engines shut down and we drifted helplessly in the North Atlantic. We radioed for assistance. An

American ship came upon us and offered to tow us into Queenstown, Ireland. We were towed for eight days and had no idea how long we'd be stuck in that place. The boiler makers worked only three hours a day and out of the three hours, they loafed around half of the time. God only knew when we'd get back to sea. I would have done anything to get out of there but there was no place to go. They had only one picture show, one eatery and one dog in that part of Ireland. There was nothing to do to kill the time. One Sunday, a couple of chaps and I went to Blarney Castle to take a look around and all we found was an old castle in ruins and a couple of caves. We quickly tired of that and went back to the ship. I must admit, though, the scenery was beautiful with its mountains and lakes. The shipyard workers in Ireland were just the slowest people. Then, finally the day came, all was ready and we set sail for Cardiff arriving there in December, some four months after setting out. Cardiff is in Wales, in case you didn't remember."

Gus leaned over the table and asked me if I would pass the jar at the other end of the table, the butt jar. "Oh no," I thought to myself. Gus took a couple of old bent butts and began peeling them. Some had brown stains on the paper. Gross, I thought. Sprinkling the leftover tobacco onto a little paper, a quick lick with his tongue and into the 'Raleigh' machine, one pull and out rolled a perfectly rolled cigarette. He tapped one end of the butt on the table and with a strike of a match, he lit up. The smoke quickly settled under the low ceilings in the kitchen. Boy, what a stench!

Back to the atlas, Gus flipped through the pages. He began again, "Now where was I?" His finger found the Atlantic Ocean and then the east coast of the United States. "Here it is," he continued, "I signed on to the *S.S. 'Lord*

Downshire' an Irish ship from Belfast. She was three thousand one hundred twelve tons."

"I remember that voyage well, rough seas, very cold weather, snow and ice, blinding conditions and icebergs. We listened carefully to the radio transmissions for iceberg sightings and distress calls. At that time the distress call was 'CDQ' in Morse Code but it was changed later to 'SOS'."

"We navigated through this area of the Atlantic, made a few iceberg sightings but managed to steer clear. We had to be on the watch for ice fields because it was hard to navigate through them. You know when you spot an iceberg what you see is not all of it. A big part of it can be submerged and cause trouble even if you take a wide berth. That's how the H.M.S. '*Titanic*' was sunk, remember?"

"We sailed all around the Mediterranean taking small cargo from one port to another. Let's see. Yes, the *'Oxonian'*, a much bigger vessel, weighing about four thousand tons…I signed on as able-body seaman but was promoted to quartermaster during the voyage."

"What's a quartermaster?" I interrupted. "I am getting to that, mate. The quartermaster has charge of the steering of the ship and all jobs connected with it, such as making and reading signals and keeping track of the hourly recording of weather. Also, you have to be able to recognize on sight, the flags and ensigns of all nations and all the various signals and codes. Remember old Johnny, way back when I first signed on a sailing ship? He once told me, 'There is no better place where a man's ambition to improve himself is sooner recognized and rewarded than in the maritime service. Initiative is rewarded, everything you do is observable.' Back to the quartermaster. Let me tell you what I know about signaling and how important this job on board is."

"Signaling is an important responsibility. It is the ability to communicate with other vessels and the shore. In 1901, the International Code of Signals was put out so as to be used by ships of most nations so we could all understand each other. Everybody then had a common language."

"The code contains instructions on how to send messages, to answer signals, a spelling table so that everyone knows the letters to use and codes for different navigation and geographic messages. There is even a specific list of 'urgent and important' types of messages that can be sent. This is how this works mate. These 'urgent and important' messages use two flags so someone observing a two flag hoist would know that this is important. Geographic messages pertaining to things like longitude, latitude and location use four flags. All other messages use three flags."

"It is important that you learn these codes by heart because there is no time to dilly-dally in looking up codes especially when there is a problem at sea."

"The code uses certain combinations of letters like Y E, which means need assistance or F H, send a boat. A skipper can communicate with vessels from other countries even when their languages are different. The messages using flags which are square and pennants which are triangular, are hoisted aloft so they can be observed visually by an observer. However, weather conditions can make it hard to see the flags so this method is not always practical. So, just as you are thinking, there are other ways of signaling."

"This other way is the use of cones, a triangular shaped object which can be set pointing up or down, a round drum shape and a ball which are fixed in order on a line and hoisted aloft. Sometimes flags or pennants are added."

"Another method is the hand flag semaphore or a fixed semaphore machine attached to the deck or fixed on coastal signaling stations. But again if the wind is blowing or the weather is not favorable these would be hard to see as well."

"So depending on conditions, the ball and cone would be hoisted up in a predetermined coded sequence, such as an upside cone followed by a drum and then a ball. This would mean, keep a lookout as it is reported that the enemy is disguised as a merchant ship. Ya get the idea mate."

"Another signaling system is the semaphore alphabet which can be sent by hand flags or a machine. The alphabet can also be sent using lights, sounds like a steam whistle, fog horn or even gun shots. The light system uses lanterns of red or white placed in certain sequences. I can give you the chart I carried in my log book as a guide for you to learn as I did."

"Another code we used was called the 'wig-wag' and the alphabet letters are in the form of the numbers (1) or (2). The message can be sent with a hand flag, a lantern, electric light or even your hat. An example of this would be like for the letter A; you would use the code (2) followed by another (2). The letter B would be (2) (1) (1) (2). The sender makes a motion like an arc to the right of his body with whatever he was using, a light, flag or his hat and if the code for the letter is (1) he motions to the right of his body. If it is a (2) then his motion is to his left. He can also motion up and down in front of his body and that would mean (3) which is mostly used for commands like a single (3) meaning end of word or (3) (3) (3) meaning end of message."

"Perhaps you've heard of another code used more today because of telegraphs and that is the 'Morse Signal Code' which also can be sent with lights or by sound. By the way mate, just so ya know sail ships use fog horns and steamers use steam whistles. Well anyway, the morse system uses dots or shorts and dashes or longs. Using a light or sound you would flash or blast a short (usually one second long) or a long (usually three seconds long) with the space between letters one second long. In addition to the letters of the alphabet there are special morse signals which can be sent without going through each letter. An example would be like a short-short-long meaning, you are standing in danger, or short-long-long meaning have encountered ice or short-short-long-short, am disabled."

"Sometimes flares or rockets are used as distress signals. They can be seen for great distances at night even miles depending on atmospheric conditions."

"Another kind of signaling is from coastal stations which are located all around the world. They use a system of flags that tell of certain weather conditions like wind, direction and hurricanes. Two red flags, one above the other means hurricane. A red flag with a white pennant means NW winds. A red flag above a white pennant is SW winds. A red pennant above a red flag meaning a NE wind and a red flag above a red pennant, SE wind. An all white flag means fair weather, a square black or blue flag signals rain or snow."

"Lights are also used on the vessels themselves on the mast head and on the ships sides. A green light would be on the starboard side and on the port side would be a red light. If you saw this at night you would be able to tell its position and direction headed. We also used sound signals during certain weather conditions where visibility was a problem: snow, heavy rain, and fog."

"The sound of a steam whistle or fog horn blasted at intervals of usually two seconds is used to warn other vessels of your presence. When two steamers pass, one short blast means my course will be to starboard while two short blasts means to port. Three blasts means, engines going at full speed astern."

"So, mate, you can see how important good communication is and the responsibility the skipper has when he puts out to sea. Here are some charts I carried with me during my sea going years. Learn 'em. Ya can never tell when

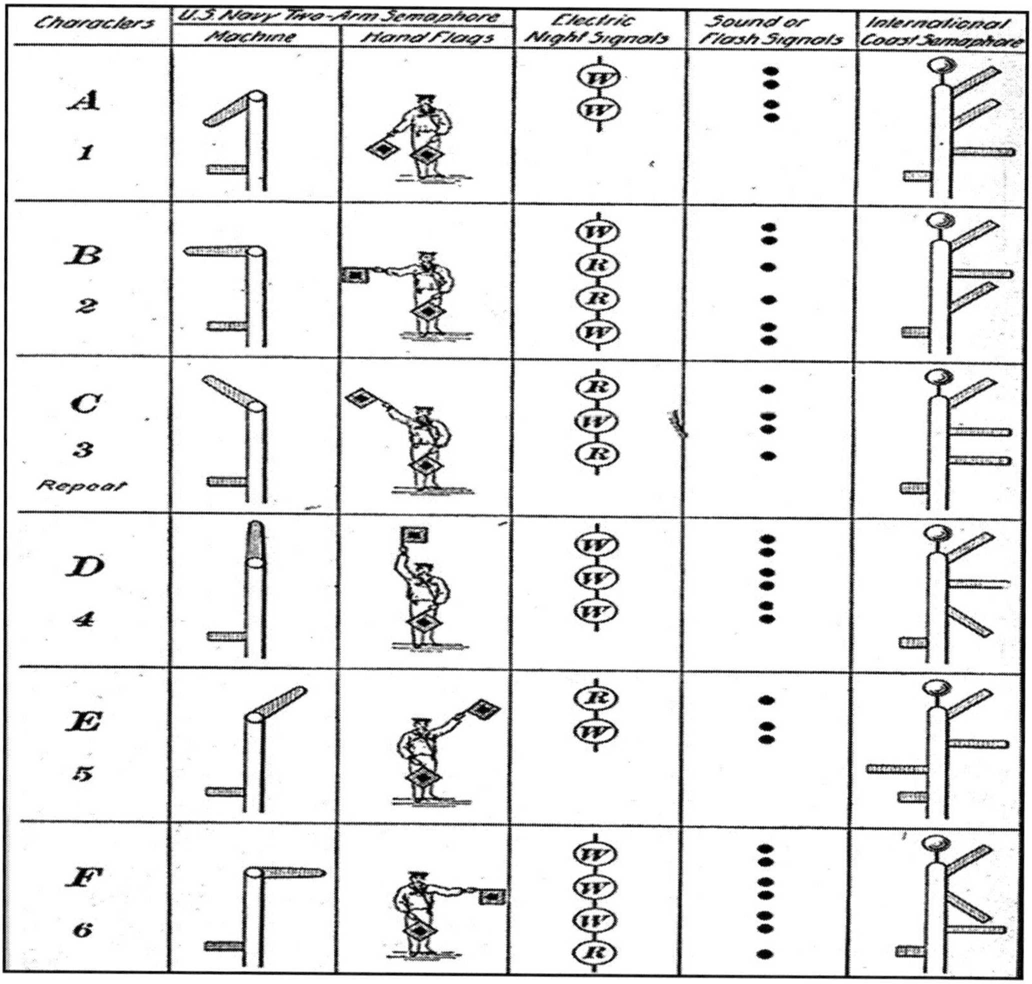

Characters	U.S. Navy Two-Arm Semaphore: Machine	U.S. Navy Two-Arm Semaphore: Hand Flags	Electric Night Signals	Sound or Flash Signals	International Coast Semaphore
G 7	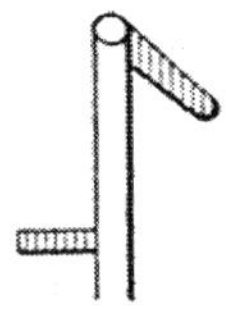		W W R R	•• •• • •	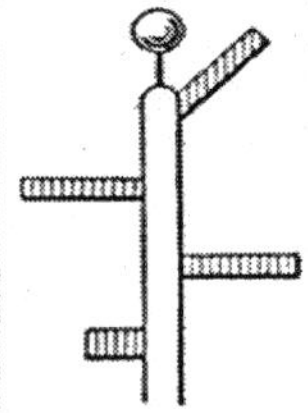
H 8	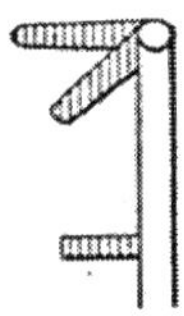		R W W	• •• ••	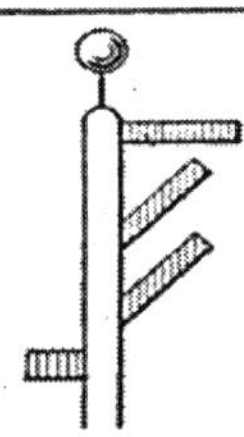
I 9	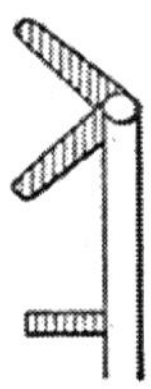		R	•	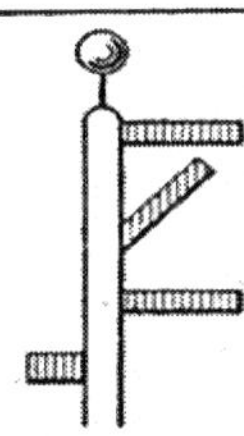
J	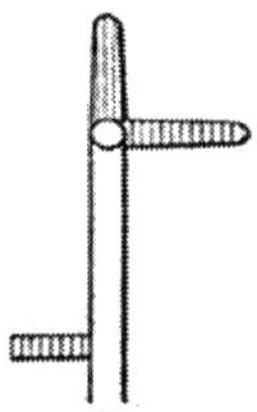		R R W W	• • •• ••	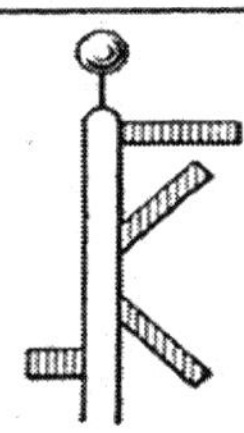
K 0 *Negative*	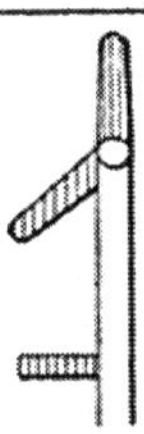		W R W R	•• • •• •	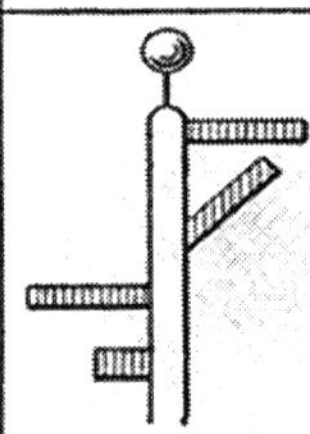
L	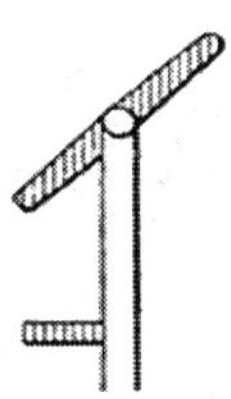		W W R	•••• •	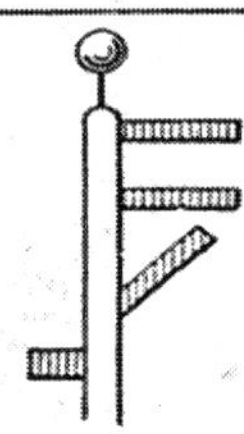

you may need them.

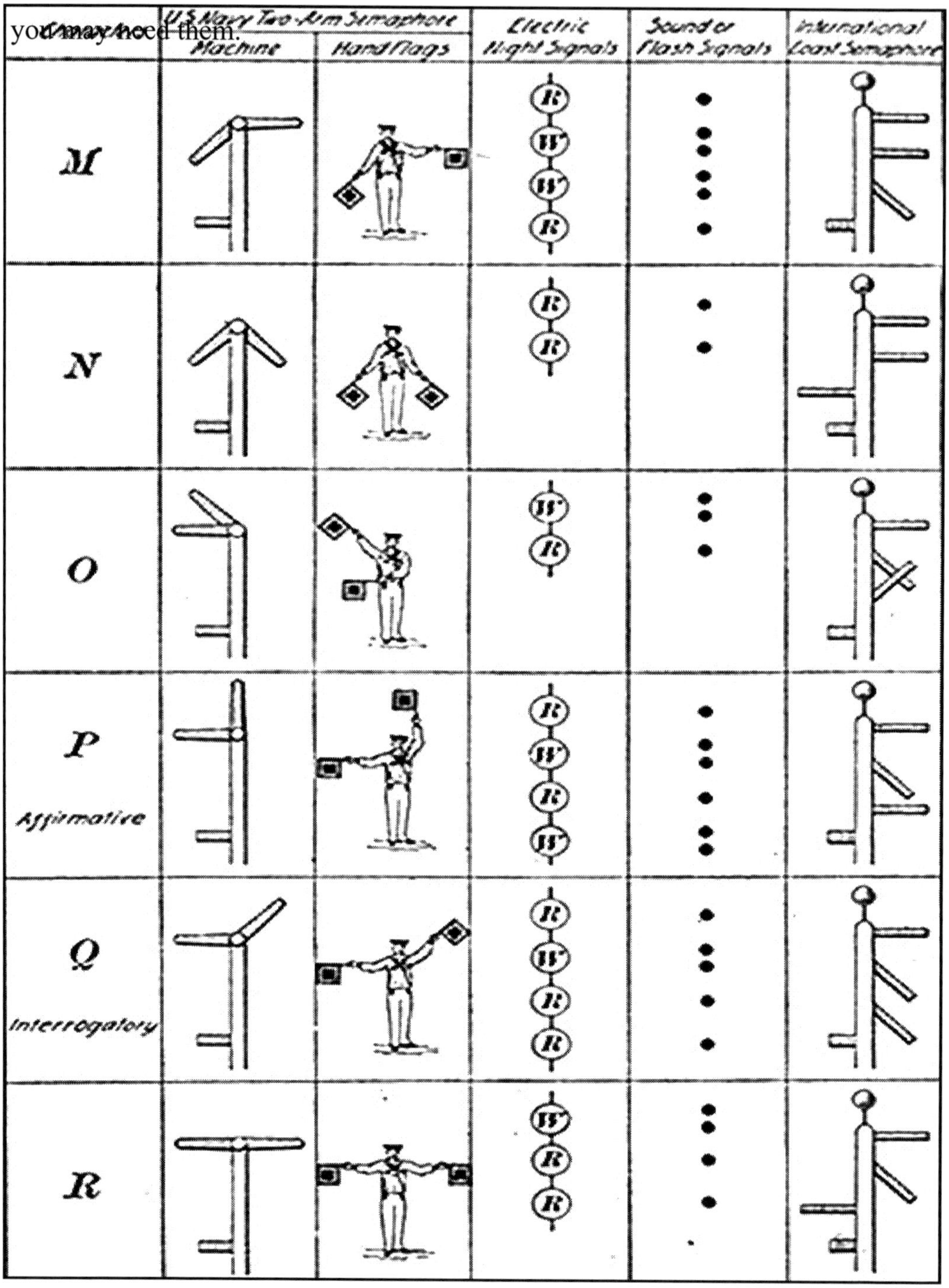

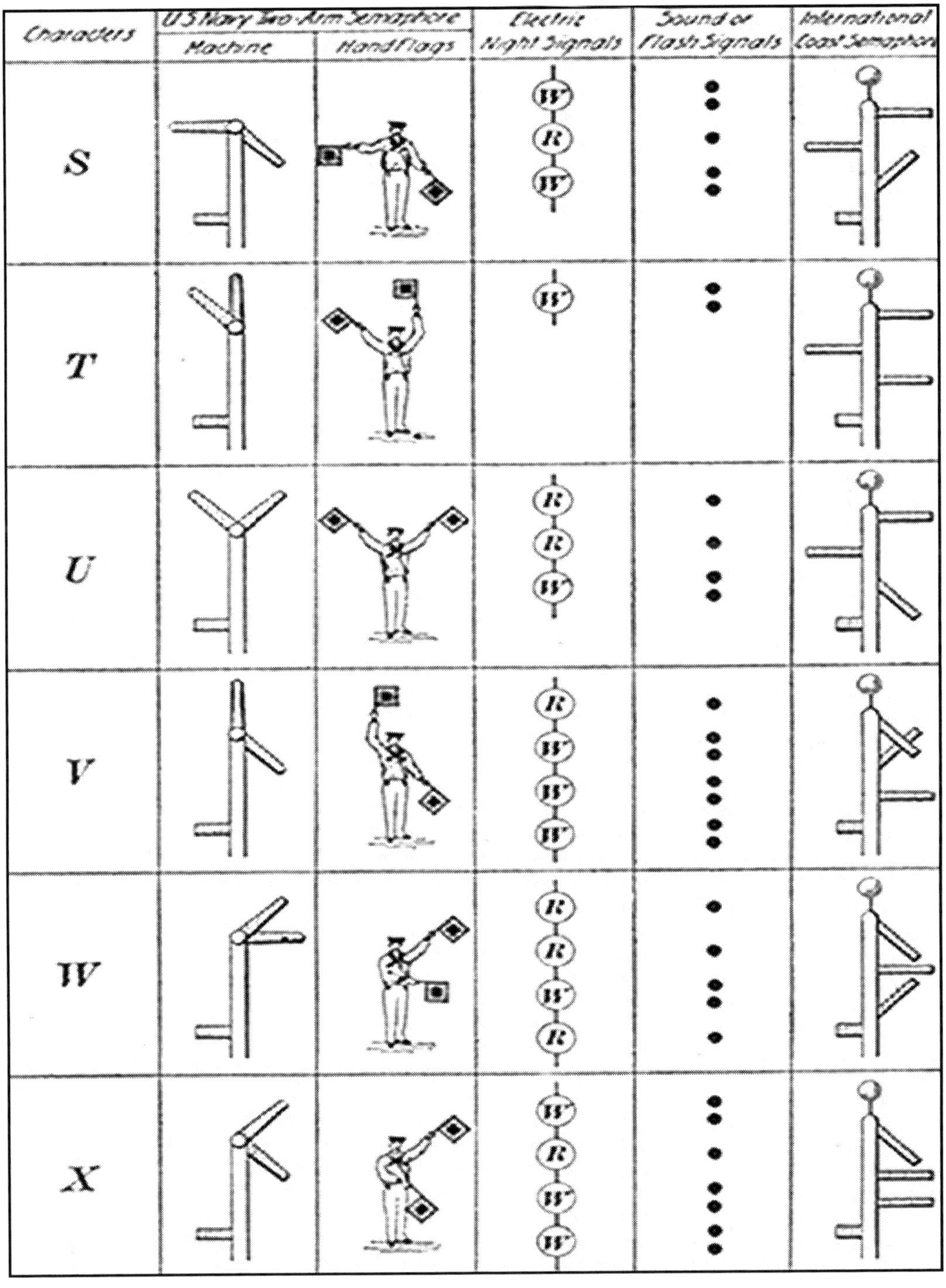
Characters
U.S. Navy Two-Arm Semaphore
Machine
Hand Flags
Electric Night Signals
Sound or Flash Signals
International Coast Semaphore
S
T
U
V
W
X

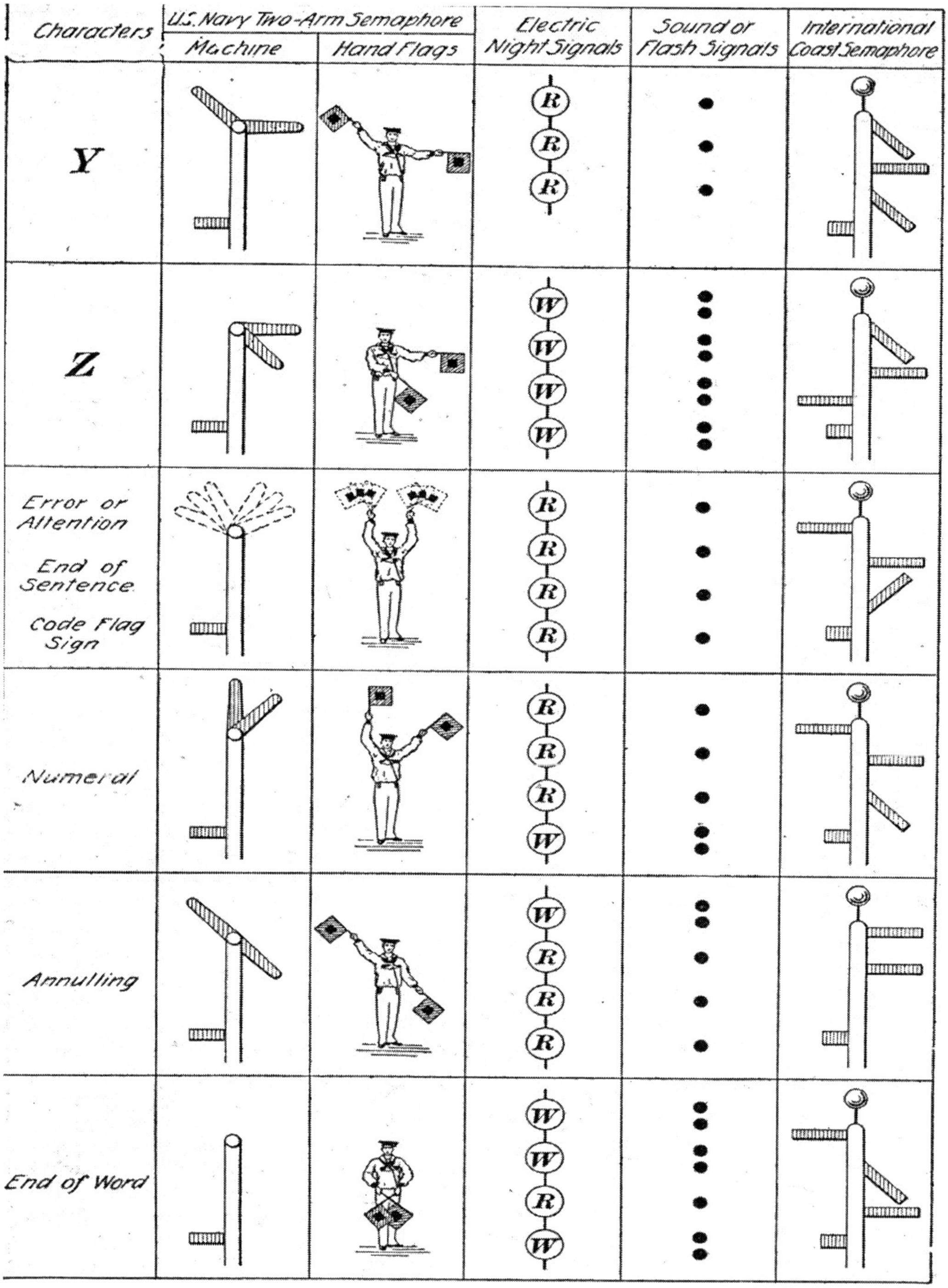
Characters
U.S. Navy Two-Arm Semaphore
Machine
Hand Flags
Electric Night Signals
Sound or Flash Signals
International Coast Semaphore
Y
R R R
Z
W W W W
Error or Attention
End of Sentence
Code Flag Sign
R R R R
Numeral
R R R W
Annulling
W R R R
End of Word
W W R W

SIGNS USED WITH THE MORSE CODE.

SIGNALLING RULES.

A *dash* is three times the length of a *dot*. The interval between the component parts of a sign is equal in duration to one dot ; that between each complete sign forming a word or group to one dash or three dots ; and that between words or groups to two dashes or six dots.

ALPHABET.

A	• —	**N**	— •
B	— • • •	**O**	— — —
C	— • — •	**P**	• — — •
D	— • •	**Q**	— — • —
E	•	**R**	• — •
F	• • — •	**S**	• • •
G	— — •	**T**	—
H	• • • •	**U**	• • —
I	• •	**V**	• • • —
J	• — — —	**W**	• — —
K	— • —	**X**	— • • —
L	• — • •	**Y**	— • — —
M	— —	**Z**	— — • •

NUMERALS.

1	• — — — —	**6**	— • • • •
2	• • — — —	**7**	— — • • •
3	• • • — —	**8**	— — — • •
4	• • • • —	**9**	— — — — •
5	• • • • •	**0**	— — — — —

SPECIAL SIGNS.

General Call	• • • • • • • • • • • •	**E's**
Answering Sign	— • — • — • — •	**N's**
Preparative	— — — •	**OE**
Repeat	• • — — • •	**IMI**
Erase	• — • — • — • —	**A's**
Understand	• • • — •	**VE**
Affirmative	• — • • — •	**AF**
Negative	— • — — —	**NO**
Numeral Sign	• • — • • •	**FI**
Spelling Sign	• • — • • • — •	**FF**
Urgent Sign	— — — — — —	**O's**
General Stop	— — — — — —	**T's**

"Back to the *'Oxonian'. We* sailed from Victoria Docks bound for New Orleans and the voyage was expected to take eight weeks. It was during winter so it was rough and cold. Ah, the *S.S 'Waiwera'*. Yes, here we are, we sailed to New Zealand, a voyage of some five and a half months. I remember the Pacific Ocean was always so beautiful maybe because the water and climate were so much warmer."

I could see Gus was getting tired and figured I'd better get myself home. I asked as I rose from my chair, "Gus, maybe we could do more tomorrow?" He replied "Sure thing, mate." Off I went home that evening thinking what great stories and what an adventurous life Gus has had. My head was filled with the imagery of all these places and of all the great oceans.

Many of the stories Gus told me were very helpful in school, especially in geography. I could locate many foreign lands, knew all the oceans intimately and could associate stories with them. Each visit with Gus followed the same routine with the traditional arm wrestle, the atlas and now the little blue book always finding its way to the table, ready and willing to divulge its great adventures under the magical kerosene lantern.

The next evening, following our traditional arm wrestle, Gus opened the little blue book and found the next entry which was the *S.S. 'Demosthenes'*.

Gus began his story, "This was a much larger vessel registered at seven thousand tons and having an engine that could make fifteen hundred horsepower. She was a transport vessel especially fitted to carry passengers and we sailed from Victoria Docks bound for Australia. I made this voyage five times and was promoted on the first voyage to lamp trimmer. On my third voyage, I was again promoted to boatswain. I know, before you even ask. What is a boatswain? This is a special job onboard. I told you about the tough boatswain I served under when I first went to sea."

"The boatswain or 'bos'n' is an assistant to the officer on deck and supervises all the work onboard the ship; all the orders are repeated by the bos'n and he has to make sure not only they are heard, but also carried out. The bos'n carries a whistle around his neck and issues commands using different calls like a bugler in the Army."

"For instance, if a boat has to be hoisted, the bos'n passes the word and then has to make sure that it is properly hooked on, that the deck crew is manning the falls and that the boat is safely hoisted and secured."

I asked Gus, "What kind of work did you have to supervise onboard?"

"Well, mate, usually we rose at 5:30 a.m., unless you had a night watch then you stayed in your rack. You get about thirty minutes to stow your hammock, maybe have a smoke and a 'java'. Then, it's 'turn to', work began with cleaning the ship as the first duty. About 6:30, if we were at anchor in port, the captain would send a boat ashore to get provisions. At 7:00 all hands turned out for mess. Mess is chow. Following that, the ensign was hoisted and it's back to work. Some crew would be assigned to clean the bright work, brass fixtures and steel and at 9:00 we all finished up, having made her 'ship shape'. Clothes lines, laundry and all cleaning gear were stowed away and the decks would gleam. Then, at noon, it was time for dinner. At 1:00 p.m. we all again turned out for work, drills and whatever the captain needed us to do. When we were done, usually 4:30, the decks were swept clean and we hung around for supper. Some chaps leant over the rail and had a smoke or wrote letters home. At sunset the ensign was lowered, boats were hoisted and secured for the night. At 7:30, it's 'pipe down' and hammocks were slung. At 8:00, the ship was inspected to be sure everything was in order and secured down. At 9:00 p.m., we usually turned in except for the crew that was on watch."

"That was pretty early to go to bed, Gus, wasn't it?" I asked. "Well, mate, it was a long day and we were pretty tired by that time. Sometimes we had to stand a watch at night and we would need to make up for lost sleep."

"What's a watch?" I asked, not quite following along.

"A day onboard ship is divided into watches: four hours each. The afternoon watch starts at 12:00 and ends at 4:00 p.m. then in the afternoon from 4:00 to 8:00 there are two watches called dog watches. This is done to make an odd number of watches so that the starboard and port watches will not be on duty at the same time every day. Then the first watch starts at 8:00 to midnight, the mid watch from midnight to 4:00 a.m., the morning watch is from 4:00 to 8:00

a.m. and finally the forenoon watch from 8:00 a.m. to 12:00 noon."

"If you are wondering how we keep time aboard ship, try to follow this. Time is kept by a system of bells. Counting from 12 o'clock noon which is 8 bells, the half hours are rung from one to eight bells which takes you to 4:00 p.m. Starting at 4:30 again, one to eight bells until 8:00 p.m. and so on. If you can follow that, you should know what watches are and how the time is known."

"Getting back to the *S.S. 'Demosthenes'*, I told you we were carrying passengers. The government was transporting people to Australia in an attempt to populate and develop that territory. Thousands of men and women were transported on ships like this one over the next few years and came from all walks of life and from all over the British Isles. Australia, as you can see here on the map, is a huge territory and has many natural resources, including gold, jewels and excellent farming lands. I heard tales about seeing jewels like opals, just lying on the ground waiting to get picked up although, I never saw any. You probably had to go into the outback to find them. I was staying away from that as I knew it was pretty rough country where one might encounter big poisonous snakes and primitive natives."

"Australia at one time was a penal colony and was used by the British for centuries. Once you were sent there you never left."

"Anyway, these voyages were far different than any I had taken before. The people we were transporting on the *S.S. 'Demosthenes'* were prisoners. The British government was emptying out its jails and sending these inmates to populate and develop Australia. It was a real problem at the time. The jails were overcrowded and most of these people were petty thieves. In those days, the smallest infraction of the law, even stealing an apple, could land you in jail.

"So the inmates were given this opportunity at a new life in Australia rather than serving the time in a British prison with a long sentence."

"As bos'n, I had the important job to make sure all the work got done to keep the ship going. It was a long journey and with all these people onboard, it certainly wasn't routine. This policy was continued by the government until the start of the First World War."

"These people were rarely allowed to come on deck and some prisoners who tried to start a mutiny or riot were handily dealt with. Well anyway, we did our job and let the government chaps guarding them do theirs."

"These ships were hired by the Admiralty for British Missions but were all manned by merchant crews. We had little training in the use of small arms; weapons were kept on hand just in case we had to use force to control the situation. Being confined, especially in the heat of the Pacific and the crowded conditions made it difficult for them and every chance we got, we opened the holds and covered them with nets to keep fresh air circulating, poor blokes!"

"T'was a long journey and I was happy to be discharged from the ship because it grated on my nerves, taking people to a place with them not knowing what to expect." Gus reached over and turned down the kerosene lamp. He was getting tired and had to be up early. "Come by tomorrow, mate, and we'll go on." Gus assured me.

In later years, my dad gave me my grandfather's papers from his service in the merchant marine and the U.S. Navy. In looking them over, I couldn't help but notice the similarity of his sailing experience with those of Gus's. My grandfather was a career sailor too. He rose up from the ranks of seaman to ship's master. During World War I, he was commissioned a lieutenant in the U.S. Navy and just prior to the war's end, he was the captain of a troop transport, the *U.S.S. 'Lake Yahara'*. Before the war, he served aboard merchant ships and made many of the same voyages to various ports as Gus did. After the war, my grandfather returned to merchant shipping and was the skipper of the *S.S. 'Columbus'* a ship of the Columbus Lines that sailed to South America and European ports of call. He was very fond of parrots and my grandmother always had several in her home while my mom was growing up.

When I was fourteen or so, about five years after my grandfather passed away, I bought my grandma a parrot and a cage. I gave it to her as a gift, hoping it would keep her company. When I spoke to her on the telephone, there was always a racket in the background. When I asked her about it she told me that it was Bobby, the parrot. He wasn't yelling but was speaking Spanish. After she passed away, we brought Bobby to Brooklyn but could only keep him for a few months as he was always yelling and laughing. The laughing sounded just like

Grandma was in the room. In warm weather my mom would put him in the kitchen window and when the seminary students and nuns marched around the block during their recess, Bobby would shriek out the window and scare the daylights out of the nuns and students. So off he went to Brooklyn Zoo where he joined other birds in the bird room. I did get to visit him years later and when I called out his name over he flew and started laughing. It was an eerie sound like Grandma was in the parrot. For all I know Bobby is still alive and entertaining visitors with his laugh today.

5

Gus Goes To War

"In August 1913, I was discharged from the *T.S.S. 'Demosthenes'* and two weeks later signed on the *T.S.S. 'Lake Michigan'* another transport. I had to take a demotion from bos'n back to able body rating as the crew already had a bos'n. We sailed from London over to Montreal, not a very long voyage, taking us a little less than six weeks round trip."

“We were carrying troops and supplies. Great Britain was preparing for war and was bringing over large contingents of troops from Canada and other parts of the kingdom.”

“The Germans had been building up their armaments for quite a few years and it was obvious they were planning to take all of Europe. There was this war fever going on and the Germans were just itching for a fight. The British had an idea this was coming and had an expeditionary force ready for deployment. If they could take Constantinople, a long war could be averted but the military people failed to give proper timing and support to the naval and military attacks on the Dardanelles and the opportunity to shorten the war was lost. The price our blokes paid amounted to millions of men being sent to the western front and the blockade of Germany by our navy, as well as the starvation of many of their people. The United States finally came into the war because it would not tolerate the sinking of its own ships. The Brits desperately needed to command the sea lanes as that was her source of food. As a result of the war, we lost over one million men killed and two million wounded.”

"That early battle on the Dardanelles was our only chance to end this mess quickly. It was the single most strategic point to put a stop to this developing conflict. There was the Suez Canal, the Persian Gulf and all that oil at stake. So, that's what the Dardanelles was about but it failed with great loss. Well anyway, I made nine voyages with the *'Lake Michigan'* as an able-body seaman. I was promoted on my tenth voyage to bos'n and continued in that position for eight more voyages."

"The *T.S.S. 'Lake Michigan'* was a Canadian registered vessel, a little smaller that the 'Demosthenes' weighing about fifty-three tons. She made nine hundred horsepower. She was a little slower than the *'Demosthenes'* as well. It was obvious that more fighting would break out at any time. Ships were being torpedoed and more German naval vessels were being spotted in the Atlantic and the Baltic. Troops were still being brought in from Australia and New Zealand."

Troops of the 2nd Gordons and 2nd Scots Guards on board the ***T.S.S. 'Lake Michigan'*** On 6th October 1914 en route to Zeebrugge

"The British Admiralty had commandeered our ships and she was now a British troop ship. The *'Lake Michigan'* could carry about two thousand troops, about two battalions. The decks were cluttered with their supplies and we had hardly any room to move around without asking someone to step aside. We traveled in convoys with a great number of vessels."

"In our convoys, the British Navy provided destroyer escorts as the German submarine was an increasing menace. It was one continuous stream of troops and supplies. We made several trips through the Suez Canal. The Suez was about eighty miles long, and it usually took us twelve or so hours to sail through. There were thousands of troops encamped on the banks of the canal… troops from India, South Africa and Brits from all the colonies."

"Port Said, the busy port I had traveled through years before, was now congested with destroyers, cruisers and, above in the skies, those new fangled air 'o' planes. From Port Said to Alexandria was another ten hour voyage down the Mediterranean. From all the activity we knew something was going to happen soon. A big battle was in the making but just where in this part of the world, we weren't sure. On one particular voyage from Cardiff, Wales, we were carrying supplies and 'munitions to Alexandria. I remember the scuttlebutt that his Royal Highness Prince Albert, the son of King George, was going to be in Alexandria at the same time we were arriving but we never did get to see him. We shipped out of Alexandria with a brigade of Australian troops and were accompanied by the *T.S.S. 'Magdalin',* also carrying Aussie troops. You could see from our stern the long column of ships in this convoy."

"We sailed out on the trade route where I had been on many times before and were joined by two escort ships. One was a minelayer and the other a destroyer both zigzagging in the waters to prevent German U-boats getting in too close. The U-boats had developed a reputation in the Atlantic for not caring whether they torpedoed a ship carrying troops, a hospital ship carrying wounded or a passenger liner. The armed escort broke off as we headed into open waters and we were now on our own. It was scary to think that somewhere beneath the surface there could be a U-boat skipper having us fixed in his scope and ready to press the 'fire' button. The captain passed the word around that he received a signal from the escort ship that the *S.S. 'Minneapolis'* was torpedoed twenty miles behind us. It was on the same route we just passed over."

"We had been trooping to France when we received new orders to steer to the eastern Mediterranean, Gallipoli. We laid over in Alexandria and had a spell with the heat and flies. Huge fleets of warships and transports now gathered and embarked from this port. There were at least a dozen battleships, including the *H.M.S. 'Queen Elizabeth', H.M.S. 'Prince George'* and even French battleships. Several ships had hit mines and exploded; their twisted hulls lay belly up in the waters. More troops were needed as this was not going to be a naval action, despite the number of ships assembled. All told, we had carried over about eighty-thousand soldiers. The British were about to make the initial assault on the coast of Turkey at Gallipoli, near the Dardanelles and the entrance to the Sea of Marmora."

British Troops ashore at Sulva Bay

"The assault began on April 15, 1915, with our chaps going ashore at different points along the coast. The going was hard fought all along and by August the casualties were very high; sometimes more than half of the men in a unit were lost with little or no ground gained. It was a futile effort, to say the least, a total waste. We sailed back and forth to Alexandria bringing in more troops and supplies. On the way out, we carried the wounded to hospital ships and then we went back to Alexandria to bring in more troops. All together, there now were about five hundred thousand troops who had landed. Our boys suffered over two hundred thousand casualties with over one-third of the Aussie and New Zealanders killed or wounded. The Aussies and the New Zealanders always seemed to get the worst of it, even over in France as the war progressed. The entire war seemed ill-fated. Our reinforcements were always too late. On top of that, the heat and the flies were terrible."

"From our decks, we could see the Turks' trenches up on the slopes and could even see them moving around. The Turks would make one assault after another and we could see them running down the slopes and heaving bombs into our chaps' trenches. Our boys would counterattack and the enemy would retreat but they always came back again; there were that many of them. The situation became so desperate that our crew and other crews had to unload the ammunition and take it ashore and up the beaches to trenches. Two crewmen to each ammo box scurried up the slopes slipping over large crags and rocks. Wounded and dead comrades were laying everywhere. There were deep, very steep gullies. Thick brush made it difficult to maneuver through. The Turks were able to fire down at us making it very difficult to advance. The bay was congested with vessels. There were two hundred merchant ships and troop carriers along with the bombardment fleet that stood in ANZAC cove and Cape Helles."

"We had no boots, only our deck shoes. It was unbelievable. Machine guns raked thousands of rounds down at us. Our boys were happy to see us;

their ammo was running low and at any time they'd have to sustain another onslaught. Back and forth, we hustled to the life boats carrying the supplies. Box after box we carried, keeping our heads low and dancing round the shale and rocks."

"Unless you were in a trench you were a perfect target; we were the only game in town. The Turks did their best to try and pick us off, but we just kept on coming and if one of us dropped, there was another crew member ready and willing to go up. It was slaughter."

"By mid-December 1915, we were ordered to evacuate the troops. As they pulled back, ammo stores were exploded sending flames and smoke into the sky that could be seen in Sulva Bay. We were taking the troops to Greece and Egypt, total mayhem."

Gus paused, apparently reflecting. Then he continued, "I made eighteen voyages on the *T.S.S. 'Lake Michigan'* from September 10, 1913 through April 16, 1918, which turned out to be my unlucky day. Sailing from Liverpool to St. John, New Brunswick, about three days out, our convoy picked up signals that German submarines were operating in the area. We took defensive positions by widening our distances, keeping in assigned columns and trying not to lag or stray off. The convoy only did about six knots; you could walk faster than that. The destroyers were running up and down our flanks. We were very anxious, after all we had been lucky for so long. The German U-boats had sunk several thousand ships already. We just hoped it wasn't our turn. We stood at the rails; because if a torpedo was fired we could see it trailing in the water. Lookouts were posted in the crows nest with binoculars scanning the surface for a wake or even a trail of a periscope. Our ship was equipped with a deck gun which would be of little help because it was small caliber, no match against a German deck gun."

"You know, I made eighteen voyages on the *'Lake Michigan'*, five years of my sea career. I grew fond of her and felt she was my home. This was war and at times, very dangerous. I got a good taste of it on April 16, 1918," Gus was reflecting here with a far-eyed gaze and his chin in his hand.

Then, "Where was I? Okay. We were three days out and we were nervous because we were told that U-boats were in the area. It's different when you know for sure they're near."

"It was quiet; all we could hear was the hum of our engines and the gentle break of the water when all of a sudden the silence was broken by a call down from the crows nest 'Torpedo off the starboard!' We looked over the starboard rail and there it was, a wake trail coming toward amidships. It was one thing to spot those torpedoes, yet another to get out of its way. The first mate ordered, 'Hard to port!' as the helmsman spun the wheel round as fast as he could. We could see the torpedo coming toward the starboard bow as the ship began its turn. 'Full ahead,' the first mate yelled out as the ship veered. It looked like we could make it. We stared at the wake as the torpedo ran ahead of the bow and just missed us. That was a close call. We all breathed a sigh of relief only to hear again, another call from the lookout, 'Torpedo off port!'"

"Those German U-boat skippers were smart, now running in packs just like wolves. That's why we called them 'wolf packs'. They would concentrate their U-boats after breaking in through our defenses, you know, the destroyer screen. Once they found a hole, it was very hard to stop them. They would set us up and if a ship could steer clear of the first torpedo, there always seemed to be another U-boat on the opposite flank for the kill. This time we couldn't maneuver fast enough and the torpedo slammed into us amidships. Just below the waterline our ship shuddered and then the sound like broken glass was deafening. There was an explosion just forward of the engine room. We started to take on water fast and began to list. Thank God we were not carrying 'munitions. The horn sounded, we took our action stations and grabbed our life jackets. The skipper and first mate gave a quick look to see the extent of the damage and everyone scurried to their posts. There was no telling if we would get hit with another torpedo. We knew they were looking through their periscope and reloading, ready to fire again if they thought we wouldn't sink. The submarine crews collected tonnage flags and when they went back to their 'den', they would drape them on their fore lines."

"I was busy with the radioman breaking up the radios just in case the Germans wanted to board us. We were a transmission ship and had special gear to relay radio signals. I tore off the metal tag numbers and wound up keeping them as souvenirs. Lookie here! I'll show you." Gus pulled out a chain with three copper tags and numbers on them.

"The quartermaster hurried to the bridge and retrieved the log book and signal codes. There was utter confusion aboard because we didn't know what would happen next. The ship was listing to port pretty good and taking on water."

"We all waited anxiously for the captain's next order. Then the order came, "Abandon ship!" Our life boats would be too difficult to lower because of the angle we were on with the water, so we jumped in. The ship was sucking bow first; listing further with each gasp."

“We swam as fast as we could to keep clear of the ship, knowing she could take us under in her draft. Steam was rushing out as the ship listed further; ash was spewing from the stack onto the water. Oil began to seep out onto the water and then it ignited, sending flames creeping like little fingers across the water.”

“The North Atlantic waters were very cold and fear set upon the crew, because we didn’t know what would happen next. The air was full of smoke and ash, the visibility was very poor. Suddenly one of the crew yelled, ‘U-boat.’ We scanned the surface. It looked like a huge whale coming up and then I could make out crew scurrying out of the coning tower. They just looked around, sized up the situation and saw no need to use their deck gun. Our ship was sinking. As silently as the U-boat appeared, she slipped back into the sea leaving us to make out for ourselves.”

“We were lucky. U-boats didn’t take survivors and were known at times to machine gun crews in the water. Totally helpless, we hung on to anything we could find.”

"We had a destroyer escort and hoped they would rescue us soon. Looking around, we could see smoke and fire down the sea lane. U-boats had done a job on our convoy which was about ninety miles North by West of Eagle Island in the North Atlantic."

"I saw the first mate struggling to keep his head above water and I grabbed hold of him. He had a huge gash on his arm and had lost a lot of blood. I took off my belt, put it on his arm tight to stop the bleeding and held him up as we bobbed around in the cold water. Finally, a destroyer found us, hoisted us out of the water and later transferred us to another freighter. Once onboard the freighter, a muster was taken of the crew; several wounded and one missing, our captain… he had gone down with his ship. Here mate, lookie here!"

Gus then unfolded a letter he had tucked away in the back of his little blue book.

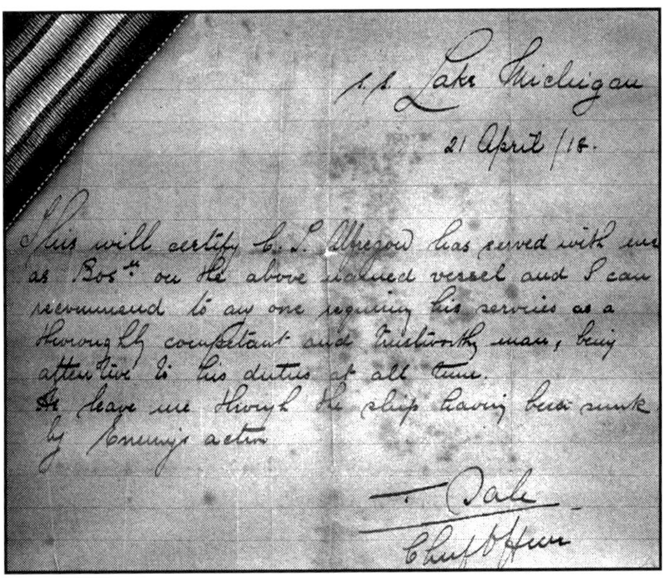

S.S. Lake Michigan

21 April /18.

This will certify C. S. [illegible] has served with me as Bos'n on the above named vessel and I can recommend to any one requiring his services as a thoroughly competent and trustworthy man, being attentive to his duties at all time.
He leave me through the ship having been sunk by enemy's action.

[illegible] Dale

Chief Officer

"My next voyage was aboard the *T.S.S. 'Lake Manitoba'* as a quartermaster from May to September, 1918, transporting troops mostly wounded from Europe back to Canada, and bringing fresh troops back to France."

"The big turnaround came as a result of the heroism of the merchant marine service, the convoy system and the naval blockade; it wore down the German people. Also we may have beaten the Germans sooner if Russia hadn't pulled out of the war in 1917 because of its revolution; this was very dangerous because the Germans were now able to bring those troops from the Russian front over to the western front. Then the U.S. came into the war but not until after the failure of the French offensive at Chemin des Dames and Ypres. We fought all autumn of 1917 in the rain and mud with little or no advance. The Germans could have negotiated a peace but they were too arrogant and underrated the Americans. The tide turned and, you know, there is an old saying, 'the English only win one battle in a war and that's the last one.'"

"Then from April 1919, I served on the *S.S. 'Burmese Prince'*, transporting troops from Havre to the United States. I served aboard her through 1921 as a bos'n and I then made several voyages on the *S.S. 'Celtic Prince'*, sailing to ports in the Far East, places like Hong Kong, Singapore and Japan. She was the last British steamer I sailed on and I was discharged in June 1922 in New York. I always had a fascination about America, having only sailed to her ports a few times. I decided to stick around and see what I could do here. I was almost thirty-seven years old, having the rating of bos'n. With all my experience, I hoped to land a good job with an American steamship company."

"It was 1922 and there was a lot going on in New York City. The war was over and the shipping business was improving. A great opportunity, I thought. I hung around the piers and eventually ran into a chap I had sailed with on the *'Lake Michigan'*. He was the first mate I helped out when we were torpedoed in the Atlantic in 1918."

"We got a room together in a boarding house in Brooklyn, right near the naval yard. It was a great place just for sailors when they came to port, mind ya; good food and plenty of fun company. The landlord was a heavy-set woman everyone affectionately called Edi. She was from Wales so we had something in common. She was quite a jovial woman, always laughing which made us sailors feel right at home."

"There I met a sailor, Edi called Nutty, a nickname for Knut, who was a regular. He was quartermaster for an oil company on the tanker *S.S 'Edward Doheny'*. He was a fun guy, very outgoing, always had something funny or sarcastic to say. Every once in a while I could tell he was tipsy, as though he had a few to many. I could never really figure him out. He referred to Edi as mother but I never found out if she was. She was in her late thirties and Nutty was in his mid-twenties, maybe."

"I met another sailor, Oscar, who was a first mate and I remember talking to him about his own experiences in World War I. Like me, he had served on troop transports in the North Atlantic and he, too, was torpedoed."

"Later he saw duty on a destroyer. We were also once in the same convoy. He was a real upright, quiet chap, always serious about everything. When he was at the boarding house he always had out his navigation books, studying, I think, for an examination for rating as a ship's master. I was never good at arithmetic, as a matter of fact, but I did always carry along an arithmetic book and practiced whenever I had the time. I just didn't have it in me to be a navigator, ya had to be a whiz at arithmetic."

"I got to know Oscar pretty well. One day he asked me if I had anything lined-up for a voyage, I told him I did not. To be perfectly honest with you, I was enjoying just sitting around and going into the city. The city was a great place, so much to do and see. I even went with Nutty to Luna Park on Coney Island and rode the steeplechase horses and went on the roller coaster. That was great. I never had so much fun since I can remember. Up on the Ferris wheel I could see out across the ocean and almost to Manhattan Island."

"Nutty didn't seem to care too much; he was always two sheets to the wind and, of course, I didn't mind as I was used to going up aloft anyway. When we got home, Oscar had some news for me. He had signed on the *S.S. 'Bird City'*, a freighter belonging to the Intern Freight Company. There was an opening for me if I wanted it but I needed to get my sailing certificate because all I had was my Board of Trade continuous certificate. I needed a U.S. certificate. So I went down to Whitehall Street, showed my credentials and they gave me an able body sailing certificate. I tried the mate's examination while I was there but it had too much arithmetic and I couldn't read English that well so I just took the seamen's license."

"Well, I could now sign on with Oscar on the *'Bird City'*. We were to sail to South America so I packed my sea bag, went down to the gangway with Oscar and settled in. The skipper was an older man who had a lot of experience on sail ships. He was an all oceans skipper and he saw this job as local work."

"We were heading for Venezuela to take on a load of bananas. This turned out to be quite an experience. Bananas come in large bunches, on stalks that had to be hand loaded. The locals did all the work. We stood by and watched as they lifted them on their shoulders and carried them up the gangway to the hold. Every once and awhile a longshoreman would yell out then step on the huge spiders that fell out between the bananas and scrambled onto the deck."

"Most of the freight we carried were fruits and once in a while some hard cargo. One voyage we brought back cattle which we stowed in the bottom holds. Boy, what a stink, hundreds of animals. We ran with the hold hatches open all the way home."

"I got to know Oscar pretty well and we were good friends. He was always considerate of me as I was always looking out for him. When we were not on duty we would talk about old times and exchange stories of how we each grew up. He was from Estonia in the Balkans, if I remember right. He also took up sailing early, came to America, became a naturalized citizen and joined the US Navy in the First War. He actually commanded his own vessel, the *S.S. 'Lake Yahara',* a troop transport. I wondered how many times our paths crossed in the night?"

"Oscar spent a lot of time in his quarters when not on the bridge. He worked on a model steamship for Edi. I had to hand it to him, he had a lot of patience; every detail was in place. I told you I made 'ships in a bottle', but Oscar's model was big. The hull he carved from a solid piece of wood, the railings and riggings were made from fine wires. On the stern, the name read *S.S. 'Edith.'* Oscar had a few peculiarities. Whenever we pulled into port, he would put on his derby and stand on the bridge as the ship slipped in. This derby was his good luck charm. I saw this with my own eyes. He dove off the bridge, disappeared under the hull and came up on the other side, still wearing his derby. Quite a feat, I must say. It always attracted the attention of those standing on the docks. Another sailor in our crew had a huge snake he kept in his quarters for good luck and would appear on the deck with it draped around his neck."

"The only good luck charms I carried around were the copper radio tags I stripped off the radios on the *'Lake Michigan'* before she sunk."

"Between voyages we just laid up in Brooklyn. Edi was a great cook and there were always plenty of goodies. She was always baking something good and the smells drifted all through the house."

"One afternoon while I was relaxing in the parlor, Edi came in and sat with me and started talking about the old times. It seemed that we had been in many of the same places and that sort of stuff. Edi had lived in Birmingham at about the same time I was there. She worked as a governess for a family."

"She came over to America in 1912 to take up a position as a governess for a wealthy New York family. She sailed from Liverpool on the *S.S. 'Celtic'*. I remembered that ship, she was steam powered but also had four auxiliary sail masts, one of the older vessels still using sail and steam."

"Edi enjoyed her job but things didn't quite work out. She also briefly talked about her husband who had been killed during the Boer War in South Africa. It was then that she set out on her own. Here she was now in Brooklyn raising her little girl, Edith."

"One thing I remember very distinctly was that on Edi's fireplace mantel were several pictures and ribbons, ya know like service medals. Edi told me her brothers had received the medals during the First War. All four were killed in combat. Edi described with great detail the names and places they fought. As I too had been involved in the war, I asked about her brothers. Her oldest brother, Harry, serving with the West Yeomanry, the Queens Own, was killed at Helles,

Turkey in 1915 in a failed assault on a Turkish gun emplacement. I couldn't help but wonder if I ever came across him on my jaunts up the slope at Gallipoli carrying ammo to the trenches back in those days."

"Edi's brother Ernest, who was a private in the first battalion of the Gloucestershire Regiment was killed in March of 1917 on the Somme in France. I also remembered that offensive, as it dragged out for years. I remember faces of the boys we transported onboard our ship and put ashore. Over seventy thousand Brits and South Africans were killed there. Then Edi's brother Charley of the 11th Battalion of the South Wales Borderers was killed in the third battle of Ypres, Belgium at a place called Messiness Ridge. Her brother Walter was in the same unit and was killed in January 1918 at Robecourt during the battle of Cambrai."

"The English lost a million men killed and many more were wounded. Although originally the idea was for England to blockade the Germans and her allies to starve them out, somewhere and some how, she got involved in the land war and, as a result, paid the price by losing a generation of England's finest. Ya know what they say back home 'The English only win one battle and that is the last one.'"

"Edi and I spent many an hour talking about England. She was a lot of fun, could make ya laugh and always held tea promptly at four in the afternoon, just like at home."

"It wasn't long before Oscar, Nutty and I were off again. But this time Oscar was the skipper. I was the bos'n and happy to have my old rating back again. Nutty was the quartermaster. This voyage took us to Valparaiso, a seaport in central Chile, west northwest of Santiago. I can't tell ya how many times I sailed through this area in the past. The sea was rough, the weather was foggy with poor visibility. Master Oscar sailed cautiously through the Panama Canal then down along the coast of South America on the Pacific side. Oscar was a confident skipper. He took frequent sightings with his sextant and was always studying his maps and making calculations. Our engineer, 'Stash' kept the engine and mechanicals in meticulous condition. He took care of the ship as if she were his mother. He always had spare parts onboard and kept her shipshape."

"Well, we picked up our cargo of fresh vegetables and headed on back to New York. The seas were calmer on the voyage back. After the *'Bird City'*, Oscar was offered the command of the *S.S. 'Columbus'* a ship of the Columbus lines out of New York. She had her own berth along side the West Side Highway. I think it was Pier 61. She was a grand vessel and could do about 15 knots. There we were, all three of us: Oscar as skipper, Nutty as quartermaster and me as bos'n. Our crews were always different. It wasn't like the old days when your crew stayed on. Now it seemed they would make a voyage, then sign on to another ship or just disappear."

"Oscar had this thing for parrots and always kept one onboard and two or three in Brooklyn. Edi was fond of parrots and whenever we returned from sea, we were always greeted by the birds with things like 'Hey Mate', 'Ship Ahoy' or 'Bring on the chow, Jack'. These were beautiful birds: blue fronts and macaws."

"Oscar, Nutty and I made quite a few voyages together on the '*Columbus*', mostly to South America. I was getting tired and restless again for more excitement. We were like a family: Edi, Oscar, Nutty and I, but I wasn't really a family man. Oscar got married to Edi and we had a little reception at the house after church. Edi's daughter, little Edith, now had herself a dad and they made for a fine family."

"I decided to return to England, to Liverpool. It just seemed at the time where I belonged. I signed onto a steamer, made the voyage back home, got an apartment and walked around looking for something to do."

"The ships were getting bigger with new fangled instruments taking the place of what I was familiar with. I was getting older and wondered if I could still fit in. I roamed around the shipyards, saw old steamers rusting in their grave yards."

"I went to the local pubs and listened to the old chaps tell war stories about voyages and adventures, each sailor trying to out do the other. I did not want to wind up like that. I was still in my mid forties and felt there was something I needed to do."

"I took a job at a shipyard as a rigger and this is where my experience going aloft paid off."

"When a ship was rigged it required men to go aloft to the mast heads to install shroud lines, antennas and the like at great distances above the decks, sometimes 100 or so feet. I had no problem with that but you'd be surprised how many chaps wouldn't do it. I had experience being hoisted in a bos'ns chair and could go above and sometimes alongside the hulls below the water line."

"I was set. I had me a little apartment right near the yard and worked there for about a year. Then guess what, I wanted to go back to sea again. I quit my job and started hanging around the docks watching the ships go in and out, just like I did when I was a kid."

Gus put aside his little blue book and said to me "That's it, mate," but assured me it was not the end of his sailing career. "I did go back to sea and I'll tell ya next time when ya come up, but ya better be off before your mother gets mad at you and me as well."

On my way down the road I thought over and over again about what Gus had told me and imagined my being a salty sea sailor standing out on a spar.

Gus was obviously not a landlubber and when he was not at sea he got antsy and just had to go back. I was beginning to understand that better. He was chasing a dream and searching for something. I wondered if he ever found it.

My dad also came to my mind and I remembered my mom telling me that when I was a year old Dad went overseas in World War II and took part in the invasion of North Africa. Now I wanted to know what he did in the war.

What ship did he sail on? Would he tell me all about the landing and combat he saw fighting the Germans and Italians.

I had some background now, a certain vested interest; I was eager to understand as much as I could and why I did not have my father for four years. Unfortunately, my dad never wanted to talk about his war time experience until I was much older.

Gus's' Little Blue Book, boatswains whistle,
radio tags on his chain and British medals:
'Service in the Merchant Marine'
Victory medal, World War 1.

6

Dry Dock

"I took some time off and just wandered around in London and New York for a while. A chap I knew heard that a crew member was needed for the yacht, *'Corsair'* which was docked down at the Manhattan piers. I went down to the pier looking for the captain. What a beautiful ship, much bigger than I thought it would be. The *'Corsair'* was owned I think by J.P. Morgan and was equipped with the latest gear which she got when taken into military service as a sub chaser in World War I."

"The captain reviewed my little blue book and discussed the duties I would be required to do. I was a lot older than most of the crew so he thought I would be an asset onboard. I was hired, went to my flat and picked up my sea bag. I was back at sea again. I met the crew and they seemed a lively bunch of blokes."

"It wasn't long until we were underway. We set sail for the French Riviera, then on to Italy and Greece. The owners entertained a lot. The *'Corsair'* was like a floating mansion. When in port elegant affairs were held on board, with the wealthiest and most royal of guests. We, of course, had to keep our place. We were the crew and had our work to attend to and couldn't engage in the socializing. We looked good though. We had uniforms that looked like those the U.S. Navy wears. Across the chest emblazoned '*Corsair.*'"

"The work aboard this ship was much different than what I was used to. Most of what I did was to keep her ship shape, everything shiny and in its place. The captain was pleased and in a few months I was promoted to bos'n. Most of the crew had little or no experience aboard large steamers. Only the captain and first officer had served in the war."

"The owners wanted experienced crew but weren't always able to find them. It was a great experience for me. Life was easy and I got to enjoy some of the more exotic places that I never had the time to see before."

"After about two years on the *'Corsair'*, I decided this was not what I was looking for. There was something I needed to do but didn't quite know what it was. I just felt I didn't fit in any more and was eager to move on. The captain treated us well and the wages were excellent, but that wasn't enough to change my mind."

“Here, mate! A picture of me when I was promoted to bos’n standing with the captain and crew. These were good easy times but not for long. The people were good to us but it got a little boring. The crews were all great, just like the good old days.”

"I spent several months jut drifting around London and then took a voyage on a tramp steamer back to New York. I took a room in Brooklyn and went to work at the Brooklyn Navy Yard as a rigger."

"This is where I met Skipper. She was looking for an experienced sailor to help her sail her fifty-two-foot cutter named the *'Nebula'*, a two-masted vessel. Her husband had just passed away and she decided to make another life for herself, pretty much what I was hoping to do. I went through the ship. She was well-fitted and looked seaworthy. Skipper and I bantered about the charter business because the *'Nebula'* was a sleek and fast ship for her class, possibly a little racing as well. It sounded exciting, perhaps this is what I now needed, some excitement again. Whatever I was looking for I didn't really know."

"Skipper is Swedish, her real name is Kirsten but everybody calls her Skipper. She came from Stockholm. Skipper had a daughter but she took ill and passed away during the great influenza epidemic of 1918, her husband passed away in 1919. All Skipper had left was the *'Nebula'* to which she devoted all her energy and time."

"We hit it off real good. I signed on and we took our first voyage to the

Caribbean just to familiarize me with the ship. The *'Nebula'* was a dream to sail. With her cutter bow she just glided through the water. All her fixtures and sails were in excellent condition; the quarters were big and roomy and could accommodate eight people. The galley was narrow as in most ships but in good shape; the galley stove was new. We had a great time and quickly took on parties of people who enjoyed sailing in the ocean and being part of the crew, so to speak. We did well enough to pay expenses and have a few bucks left over. After all, this was what I started out doing. My experience with sailing came in real handy."

"I got to know the *'Nebula'* inside and out. Skipper and I had no trouble handling the ship ourselves. We were excited with the charter business and looked forward to it growing. We took charter parties out for weekends and week long voyages. Skipper did the cooking and I pretty much handled the sail and navigation."

"The stock market crash of 1929 hurt us badly. The charter business took a nose dive and we soon drained our resources. The '*Nebula*' sat idle in dock for months. Then came Prohibition. We heard through the grape vine that there was money to be made in liquor and certain people were looking for fast ships to transport it from the Caribbean to the States."

"It was dangerous work but we needed the money to keep the ship; it was all we had. Just take on a cargo of liquor, bring it back to the U.S and make a good profit. So that's what we did. We'd run the liquor up to New York, sit in the outer limits at a given point and wait for the power boats to take the delivery."

"That's all we had to do, but of course, we had to be on the lookout for the Coast Guard. This wasn't legal and there were serious consequences if ya got caught. We could lose everything and I'm sure go to jail."

"We ran back and forth to the Caribbean with our cargo and after expenses were able to put aside a few bucks. The *'Nebula'* started to become a problem. She was built in Great Britain in 1865. Seventy years old, she was in need of serious repair. This was beyond our means. We ran liquor but not with the same vivacity. We had to handle the ship gently and kept her going as best we could."

"Well, a decision had to made about the *'Nebula'*. She was old, in dire need of repair and we could not afford to take a chance sailing her anymore."

"So Skipper decided to follow an old tradition when a ship is to be retired, bury her at sea. We took off all her fixtures, lines, brass fittings and even her bell, there it is hanging by the door, mate." 1935 ended our lives at sea. We had many discussions as we pondered what was next in life for us."

All fixtures removed the
'Nebula' is buried
at sea

7

Land Ho! Mate

"What about farming" I thought, a lot of sailors I've heard, go into farming when they retire. Skipper thought it over. Maybe if we went into a specialty business, we could make a go of it."

"How about raising turkeys, we thought. Skipper looked into it and it seemed that there was a demand for dressed turkeys."

"So we set out looking for a farm where we could start our little business. As a young boy, back in Wisby, I had helped out the local farmers and knew a little about horses and cows."

"Isn't it ironic that I left home to get away from that life and go to sea and here I was leaving the sea and becoming a farmer. So here I am mate."

"We looked around and finally found the perfect place, a big old house with plenty of room, three big barns with stanchions and stalls all on two hundred and twenty five acres."

"Part of the deal included two fine work horses, equipment to do the haying, spike–toothed harrows and a big wagon. I got right to work building turkey coops and getting the place ship-shape. Skipper got a flock of chickens so we were also in the egg business."

"Our turkeys arrived, one thousand of them. I set them in their coops and now we were in business. I made sure the coops were secure with wire because it didn't take long for the foxes and raccoons to start coming around. The 'coons' were tough because they could chew through ordinary chicken wire so I had to use something tougher."

"We bought a milk cow and named her Bessie, then a bull to keep her in milk. We named him Old Man. The pastures were all fenced with barbed wire. I put the Old Man out in the west pasture and that's where he stayed for the summer and in the winter he had a stall in the barn."

"There were times when he was hard to handle but with a ring in his nose and on a line, he behaved himself. Then there were Dick and Boy, the work horses of the farm, a beautiful pair of horses with plenty of experience in harness. They could pull a wagon, draw logs out of the woods and pull cars out of the mud."

"Farming was like sailing in a way. It had its routine that needed to be done everyday. The cow had to be milked, the animals fed and kept warm in the winter. The fields had to be taken care of, hay cut and brought in, and then there was the firewood which took all summer to collect. That's what it took to keep this place ship shape and running."

"There are other chaps too. You've seen around the barnyard, mate. There's Lymey the goat, a very smart fellow. He follows me around every chance he gets and always wants my attention. He especially wants a little rub now and then between the horns. His partner, Johnny, as you noticed, is not very friendly. He likes to give me a boot in the 'arse' with his horns whenever he can. They're all good old chaps, though."

"You've met Bos'n, my little duck. He's always checking up on me to be sure I'm doing my job right. I hear it from him when he's displeased with my work. A little handful of cracked corn and he forgets all about it. It's a lot of fun and work keeping all these buggers happy but it's worth it. Then there is Dick and Boy, the hardiest of my loyal crew. I get a lot of pleasure from my crew of guys and wouldn't give it up for anything in the world, not even to go back to sea."

Full ahead!
Mate Lymey

Clear the deck!

All hands on deck!
Bos'ns apprentices

“The turkey business didn’t do so well. It turned out a lot tougher than we thought. Many of the turkeys got sick from disease, quite a few were lost to predators. The hardest part of the job was getting them ready for the market. We had no electricity or running water so it was hard to prepare them.”

“Then the war in Europe broke out, World War II, and supplies were getting harder and harder to get. We were lucky having milk and eggs but a lot of other things were rationed and hard to get.”

Skipper tending her chickens

“I felt I had to do something so I decided to do my part and go down to the Navy Yard again and work the riggings. Skipper managed the farm and was able to hire a man to help. I came up from Brooklyn every chance I could. I did my part for almost a year and finally came back to the farm. We were barely able to get by. Skipper sold milk and eggs but that was about it. After the war we tried our hand at raising Guinea hens but again that market never panned out for us.”

“So, that’s about it, mate, the whole story,” Gus said as he reached over to the kerosene lamp and turned down the wick. It was very late, I thanked Gus for sharing all those wonderful adventurous stories with me; we exchanged hugs and down the road I went. I knew the road like the back of my hand, every bump and turn along the way. The critters and I were now long time friends.

Summers and vacations came and went. Many, many hours were spent at the table reliving the stories over and over until I had them virtually memorized. We enjoyed the afternoons together under the old ash tree, singing sea chanties until eventually I learned them also. I continued to help Gus get the hay in and do other chores as the years passed on. I was getting older and had my own new adventures to experience with college and then the Army.

Gus and Skipper didn't live alone up on the hill. If you walked down the road past their house, the road continued but had become more like a trail for lack of use. About a mile and a half through the woods lived Clara. She was very poor, had no running water or electricity and she made her living from the sales of chickens and other small livestock she raised. Clara would come up to visit Skipper every spring, have her hair cut and talk. I met Clara and she invited me to come over and see her animals. She had a black lamb she was selling and after I told my mom, she thought she'd like to buy her. The next day Skipper, Mom, my brother Henry and I, walked the mile and a half through the woods until we came to an old shack. Across the road were old car bodies with pigs, chickens and goats poised and perched in and around them. Skipper knocked and went into the house. Clara had company and we were all introduced to a lady and her daughter from Lenox who were looking to buy some chickens.

Mom said hello to Lillian and her daughter whose name was Edith, just like my mom's name. It's funny, my grandmother's name was Edith, too. I never thought there could be so many 'Ediths' in the world. Lillian and her daughter also had their eyes on the black lamb. She was a beautiful black lamb with a smooth black snout. Her wool was curly and Mom took a fancy to her right away. Edith was disappointed when she found out my mom wanted to buy the lamb. As a matter of fact, she took it out on me and was not very friendly to say the least. I wouldn't have minded a couple of little Edith's chickens and you could be sure that at that point, she definitely would not have parted with even a one.

Henry picked up the lamb and was carrying her around the barnyard so the deal was sealed. As for Lillian's daughter, who would have guessed I would meet her decades later and that some fifty years later, we would marry. Just like Gus and Skipper had found one another at a time in their lives when each was struggling to begin a new life, Edith and I met again too, at a critical time in our own lives.

That little lamb became the first of many sheep and lambs my mom raised. At one time Mom had forty ewes and every fall we would have the sheep sheared and sell their wool. At lambing time, Mom built little cubicles for the ewes in her big living room so they could be warm and safe. After Gus passed away, the sheep kept the pastures down to the roots, like a carpet.

8

Ships of Wood - Men of Iron

Gus had a way about him that was entirely unique. "There is only one way to do anything: the right way," he often told me "anything less is unacceptable." He was an early riser, a diligent worker and did not stop until the job was done even if it meant working into the night. He was totally self-disciplined, self-reliant and expected the same from everyone around him. He was a boatswain in the true sense: commanding, charismatic and strong. He was not a person to mess with. He was light on his toes and quick to anger. Most people, if not all that I knew, found him intimidating. He made no bones about how he felt and didn't make any effort to be cordial or accommodating. He was a 'strictly business' guy. He did have manners and could be polite, but it is his directness that is most remembered. Gus wasn't an angry man, just a self-sufficient, independent individual possessed with an inner strength that exuded total autonomy. He didn't need anyone. He was every inch a sailor, resourceful and obedient to rules.

Gus was agile on his feet and strong as an ox even in his early seventies. His powerful personality was reflected through his steel gray blue eyes. They seemed to reach right inside of you. He had absolutely no fear. I always felt safe with him no matter where we were and would have followed him up the tallest mast.

I remember once when I was sixteen, Gus was annoyed with me for talking back to him over something. I don't remember, but words led to an argument. He came up to me and stood in my face, steaming mad and his eyes squinting. His body stiffened up, and then he lunged at me. I managed to duck just in time and started running down the road. Looking over my shoulder, I could see him catching up to me, then just when I was at my door, he grabbed hold of me, his face now flushed. He looked me square in the eye, started grinning and then burst out in laughter. He wanted to let me know that he was still the bos'n and I should never forget that, not even for a moment.

In the spring of 1955, Mom was redecorating the house in Brooklyn and Gus agreed to come down to the city and do some painting. I came up on the bus and took him back to Brooklyn just to make sure he didn't get lost. It seemed the easiest way to get him down to the city. Gus put on his Sunday suit and his wide-brimmed hat and we set off down the road to catch the bus. Gus's suit looked vintage 1920s, all wool and solid black. It hadn't been pressed in a month of Sundays. On his head was a black fedora that had the widest brim, like I had seen in gangster movies. It must have been a nice suit in its day. This was the first time, however, I saw Gus dressed-up or even wearing a tie. He looked great. We stood by the old milk platform waiting for the bus. It finally came along, slowed down and pulled onto the shoulder. We got in and we took the seat behind the driver, as always reserved for me, and now Gus also. Gus was infatuated with the bus ride. It was probably many years since he had ventured this far from home. He spent much of the trip just gazing out the window and taking in the sights. Living up on the hill without a car, Gus did not have an opportunity to go off anywhere.

We arrived at the 34th Street Greyhound Terminal and we walked down to the subway. We had managed to arrive at the beginning of the rush hour and there were so many people. No one was polite; people were pushing and shoving. I got a little worried that Gus would get mad so I kept talking to distract him. The BMT pulled in and with good luck, the doors opened right in front of us. We hurried on and managed to get two seats before the crowd rushed in. I noticed Gus take something out of his pocket quickly and it looked like he slid it under his seat. People were rude and staring at him. I think it was his suit and hat. They were boldly giving him the once over and Gus was getting agitated. These were the days of thin ties and narrow lapels. Gus's tie was half a block wide as were the lapels on his suit, definitely not in vogue for the time. He looked like a cast member for a roaring twenties set…I would have agreed, but after all he was Gus. I was busy watching for our stop when all of a sudden, I heard a 'blat, blat.' Oh, no I thought, he brought the machine. I turned and looked at him. I noticed that whenever he saw somebody looking at him, he would glare back, squint his eyes, lift off his seat, and let go a 'blat, blat'. Each person would look away with disgust.

The man sitting next to Gus was trying to read his newspaper, but he got up and preferring to stand than to sit next to him. Finally we reached our stop and got off. Thank the Lord!

Carrying a suitcase that looked like it came off a Wells Fargo stage coach, we navigated through the crowd and walked down several streets to my house. He stood out, believe me. Everyone who passed us on the street had these funny looks on their faces. They would think twice if they saw his tattoos and his bulging muscles. Finally, we arrived at the house and Mom had coffee ready, Gus's 'java.'

After a few days, Gus was finished painting and ready to go home. I went with him to the bus terminal. Gus wanted to stop on the way at the Navy Yard just for a look. There wasn't much going on. Gus just stared out over the water, it was very nostalgic for him.

I accompanied Gus on the bus and asked the driver to let him off at Ed Jones Road. I knew that my friend would see that Gus would not fall asleep and miss the stop. Gus sat down in our reserved seat behind the driver and off the bus went. Gus would be okay, I was sure.

There were so many stories, adventures and funny situations to be told that whenever or wherever I was with Gus I always had a good time. I enjoyed

Skipper as well. She was an intelligent woman and avid reader, many times staying up through the night reading. I never saw a radio in the house or a newspaper, just magazines like 'Life' or 'Look'.

Skipper was outgoing, a good listener, always willing to lend an ear if I ever had a problem. Gus, on the other hand, demanded respect from a person before he would share anything, let alone advice. He just didn't demand respect but a kind of loyalty, more like the kind of respect and loyalty one has for someone's rank. Everything Gus did had a method and it was well thought-out. He stuck hard and fast to rules, never deviating. Men were supposed to be intelligent and obedient and he often spoke of the days "When ships were made of wood and men made of iron." He viewed many men today as "Ships made of iron and men made of wood," and believed our minds and bodies are shaped through discipline and hard work, these traits were lacking. Men have to master their world, not rely on someone else to solve their own problems, be able to use their own wits, intelligence and resources.

Gus frequently reminded me that "There was neither wind that couldn't be tamed nor sea that couldn't be sailed."

Gus derived great pleasure from his crew: Bos'n, Lymey, Dick and Boy and all his other barnyard friends that he considered as companions in his world. He was loving and considerate of them as a bos'n would be of his crew, one who truly cared for their welfare.

As time passed, many of his little friends sailed off to the 'Isle of the Blessed' as Gus put it and he himself was getting older and not able to keep up. Skipper had to bring in part time help to do the big jobs like haying and other labor intensive tasks. Boy passed away and over time there were just a few of the crew left. Gus spent hours sitting under the old ash tree, reflecting over his life and his past voyages.

His surviving crew was loyal to him and looked forward to being with him under the old ash tree. I visited as my time permitted because I was now in college. When I had the chance, I came up and sat around the table under the glow of the old kerosene light, chewed the fat, and of course, relived many of Gus's sea adventures and sang chanties with him. It was very hard for me to see such a dynamic and strong man succumb to old age. His eyes lost their glint and he began having trouble getting around.

Skipper told my mom that Gus was failing and she didn't know what to do about it. A year or so before, when my dad was up for a weekend, he examined Gus and I'm sure that if anything could have been done, he would have taken care of it. I'm also sure that Gus knew that he was sick and knew what was wrong. My dad was always up front with his patients and Gus was no exception.

Gus just sat it out like the ancient mariner he was, never complaining. He never said anything about how he felt and I didn't dare ask. There was that side of him, guarded and very private just like my grandfather. A wisdom or knowingness, a consolation, perhaps the 'secret of the sea'.

One afternoon while I was up visiting him from the city, Gus called me into his room and told me, "I want you to have my atlas and my little blue book and you take good care of them for me. I'm not planning to go back to sea but if by chance I do, I'll know where to find them." He reminded me of my grandfather in that he sat me down to share something so special with me.

Sitting beside him under the old ash tree, I would notice a certain quietude, a certain resignation. He gazed off into the horizon reminding me of Longfellow's poem, 'Secret of the Sea'. "What pleasant visions do haunt me as I gaze upon the sea..."

During the early spring of 1960, Skipper called my mom and told her that Gus was rushed to the hospital and operated on. He had stomach cancer, but in surgery, the doctors found that there wasn't anything that could be done.

Skipper related the events leading up to that moment. Gus continued to lose a lot of weight, became weaker and weaker and refused to go back to see a doctor. He was eating very little and finally just collapsed.

I went to see him and found him in his room curled up on his bunk. I had all I could do to contain myself. Gus had completely wasted away. He looked up at me, his eyes now dim and then said to me "Cheese and Rice, they wouldn't leave me be!" We sat and said little. Every now and then I could see a tear run down his face. He grasped my arm and squeezed it as the emotion swelled in his eyes and he said to me "So long, mate, remember, don't take any wooden nickels."

Gus passed away within a week. He was buried in the old cemetery a mile down the road. After his funeral, I walked through the fields and out to the edge of the woods where Gus and I often went to get firewood. I walked into the barn, now quiet. Dick was gone and Bessie too. The only chap left around the place was Gus's duck, Bos'n and where do you think he was, as usual, on watch at his post, waiting for Gus to show up and do the chores so he could have his ration of cracked corn. Everything was gone, the harness, the mowing machine, the wagon. It was just too much to take care of. Skipper had to sell them or find a home for them.

Throughout my life, Gus's old sayings have often come to mind and I have relied on this wisdom when I faced my own difficulties. I have asked myself, "What would Gus do?" I thought that maybe someday I would live in the Berkshires like Gus. I thought maybe someday I would have my own little crew but first I had to embark upon my own voyages around the 'Horn' and the over the 'rollin forties'.

Bos'n at his post standing watch

9

It's a Small World

As an older teenager I spent many hours at the dining table at home talking with my mom. She told me stories about when she was young and what it was like not having her father around. I found it interesting to hear her experience growing up in a boarding house in Brooklyn where sailors often stayed. My grandfather was often gone on sea voyages.

I listened carefully and the more she told me about her childhood, the more I remembered what Gus told me years before about his living in Brooklyn. As it turned out, he was a boarder at my grandmother's house and also sailed with my grandfather Oscar, on many voyages. I was able to share with Mom that story about Nutty. She remembered him and this led to her being able to fill in some of her own forgotten memories.

What a small world I thought. What a connection it was to make, growing up and hearing Gus's stories and later to find out that Gus and my grandfather had sailed together and that it was my grandmother who spent time with Gus conversing about old times in England.

My mom was quite surprised as I related Gus's story but I could hear in her voice a reservation about those days. Just imagine living with sailors in a boarding house, sharing dinner with them, becoming friends. Then they would leave, possibly to return in a couple or three months or maybe never at all. Having a father away at sea most of the time and having invested in these relationships only to have them vanish or become so temporary must have been disheartening for her as a child.

From the moment I met him, I was drawn to Gus. He was like an old friend I had always known. In later years came the knowledge he was a part of my mom's childhood as well. Just imagine had Gus realized that down that rural road in the Berkshires was the very same 'little Edith' that he knew briefly back in his younger days.

Epilogue

Putting this story to words was a great exercise in the recall of many dusty memories that brought to mind the smell of wood smoke, kerosene and the odor of old barns. My research into Gus's old log books revived in me the idea of being considered 'seaworthy'.

While writing this story, I also had the opportunity to reflect on the meaning of certain encounters and events in my own life's journey. As my life progresses, I have come to appreciate more and more, the value of Gus's wisdom and experiences that helped me understand the meaning of earned relationships, mutual respect and an 'eyes wide open' view of the world. Gus taught me the importance of what makes up the true measure of a man: discipline, respect, diligence, duty and integrity.

Through his photographs, log books and the many conversations imprinted in my mind, I set out to recreate those easy days when fascination and adoration of an old man and his life experiences were so important to me. Living in the world today with all that is going on, I found myself back in that safe place, amid the soft summer breezes, the shade of the old ash tree and in the company of the old man who in his own way cared about me and I about him. Unfortunately for me I was never to have that experience again. People seem so preoccupied with their own personal affairs today that they never can find the time or see the value of such sharing. A neo-cultural attitude or something, I suppose. As an educator, I respect the value of mentoring but society today is not interested in the ideas of others but rather fixed and intransigent with their own views, no matter what experience or gained wisdom another person may have.

Where are the 'Men of Iron' that characterized the generations of my parents and Gus? Not only does it appear that we have compromised our values and readily accepted those of others, through such concepts as 'consensus', 'brainstorming' and 'group decision making', we have become paralyzed by our inability to respond and act with personal responsibility, self-determination and the visceral strength that was once the hallmark of our nation's will.

Now we are in an age of fantasy with fictional heroes, glorifying those of questionable, immoral character, conveying the idea that adversity may be overcome by magic, violence or skullduggery as represented in today's literature, music and films.

Where are the real heroes? Those whose staunch individuality overcame the evils they faced, and through their persistence conquered, not by accepting pacification, or other nonsense but by standing tall in the face of it.

"Where are the Men of Iron who sailed on Wooden Ships?"

Gus's Sea Chanties & Song

'Kings Chanties'

"Aboard ship nearly every man forward has a musical ear and on every pull in calm and in storm the ship is enlivened by chanty singing. I was fortunate, for on my first voyage from Philadelphia to Japan and home again to New York, we sang chanties constantly. We had an African stewardess, the wife of our cook and at times she would come out of the galley, stand in with the men and lead off on a chanty.

"On another ship we had an old 'shell back.' His voice was a loud rich baritone. We always looked to him to lead us in a chanty. From him I learned nearly all the chanties I had heard sung for I was then young and impressionable. Later on, as I launched into manhood, I heard the same chanties sung by other sailors but they varied in tune

"A chanty is an inestimable asset to the work on a sailing ship. One sailor makes a noise when others are hauling on a rope with him and at the sound of his voice they pull together. The noise like the chanty-man singing produces team work and co-ordination. On the forecastle head, heaving on the windlass brakes or walking around the capstan, pushing on the capstan bar, the chanty-man sings the solo while his mates join in the refrain. To mast-head, a topsail yard, or any hauling of length, a long drag chanty is sung. To haul on a rope already taut, or to sweat up on a halyard, a short drag chanty is sung.

"When pumping the ship, for some 'old hookers' leaked considerably, a tail rope was put on the pump handles and then the chanty-man sang a pumping chanty. The words of most of the chanties were improvised by the chanty-man. In rhythm he sang of the virtues and failings of the captain and the crew *.....

* 'Kings Book of Chanties' Stanton H. King
Official Government Chanty-Man
Oliver Ditson Company, Boston 1918

Gus's Sea Chanties & Songs

1- **Blow Boys Blow** long drag chanty

2- **Blow the Man Down** long drag chanty

3- **Whiskey for Me Johnny** long drag chanty

4- **Hanging Johnny** long drag chanty

5- **Blow the Kaiser Down** long drag chanty

6- **The Dead Horse** long drag chanty

7- **One More Day** pumpin' chanty

8- **Leave Her, Johnny** long drag chanty

9- **Rollin' Home** old Sea Song

10- **Storm Along** pumpin' chanty

These were Gus's favorite sea chanties learned during the forty or so years he spent at sea. Gus was his own chanty-man, while sitting on the well head or under the old ash tree he always had a chanty going.

Blow Boys Blow

Long Drag Chanty

A Yankee ship came up the river
blow boys blow
a Yankee ship with a Yankee skipper
blow my bully boys blow

A Yankee ship is on the river
blow boys blow
her masts they bend, her sails they shiver
blow my bully boys blow

How do you know she's a Yankee clipper
blow boys blow
the Stars and Stripes they fly abovc hcr
blow my bully boys blow

Who do you think is Captain of her
blow boys blow
old Holy Joe, he is the skipper
blow my bully boys blow

Blow Boys Blow **(continued)**

What do you think they have for dinner
blow boys blow
hot water soup but slightly thinner
blow boys bully boys blow

Oh blow today and blow tomorrow
blow boys blow
and blow for all the salts in sorrow
blow my bully boys blow

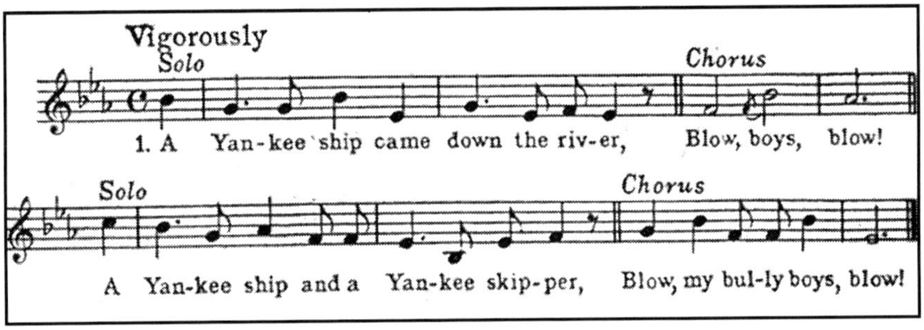

Blow the Man Down

Long Drag Chanty

Oh, blow the man down, bullies, knock him right down
Way——,blow the man down!

Oh, blow the man down, bullies
knock him right down
give me some time to
blow the man down

As I was a walking down Paradise street
Way——, blow the man down

A brass bound policeman,
I chanced for to meet
give me some time to
blow the man down

I hailed him in English
and hailed him all 'round
Way—, blow the man down

Ship ahoy, ship ahoy,
Oh, where are you bound
give me some time to
blow the man down

Blow the Man Down (continued)

A-watching the damsels
so gay and so young
Way——, blow the man down

It's arm-in-arm
we strolled'round the town
give me some time to
blow the man down

Oh, policeman, policeman
Please come along
Way——, blow the man down

I'm a flying fish sailor
just home from Hong Kong
give me some time to
blow the man down

Whisky for Me Johnny

Long drag chanty

Oh, whisky is the life of man
whis-ky John-ny
whisky is the life of man
whis-ky for me John-ny

I drink whisky when I can
whis-ky John-ny
I'll drink it out of an old tin can
whis-ky for me John-ny

Whisky gave me a broken nose
whis-ky John-ny
whis-ky made me pawn me clothes
whis-ky for me John-ny

whis-ky drove me 'round Cape Horn
whis-ky John-ny
t'was many a month when I was gone
whis-ky for me John-ny

I thought I heard the old man say
whis-ky John-ny
I'll treat me crew in a decent way
whis-ky for me John-ny

Whisky for Me Johnny (continued)

A glass of grog for every man
whis-ky John-ny
and a bottle full for the chanty-man
whis-ky for me Johnny

Whisky dried me tears away
whis-ky John-ny
whisky took all me cares away
whis-ky for me John-ny

Whisky makes me feel so sad
whis-ky for me John-ny
whisky killed me dear old dad
Whis-ky for me John-ny

Whisky took me brains away
whis-ky John-ny
one more pull and then belay
a whisky for me Johnny

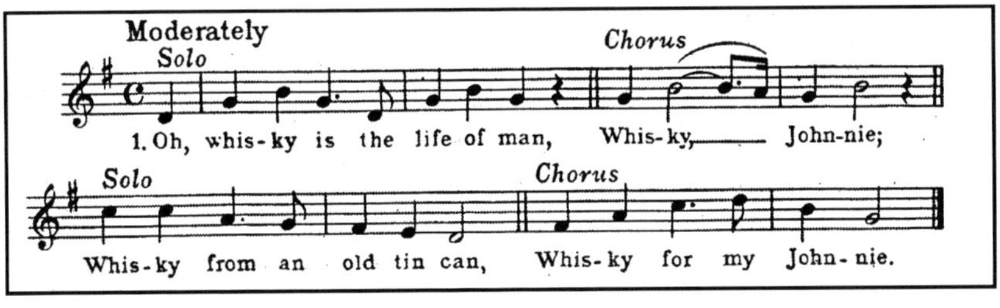

Hanging Johnny

Long drag chanty

They call me hanging Johnny
hooray, hooray
but I never hung nobody
hang boys, hang

They say I hung me mother
hooray, hooray
and then I hung me brother
hang boys, hang

They say I hung for money
hooray, hooray
they say I am so funny
hang boys, hang

They say I hung a robber
hooray, hooray
they say I hung a jobber
hang boys, hang

They call me hanging Johnny
hooray, hooray
but I never hung nobody
hang boys, hang

Hanging Johnny (continued)

They say why do you do this Mate
hooray, hooray
I know nothin' else I say
hang boys, hang

They say I hung me uncle
hooray, hooray
but that's a lot of bungkle
hang boys, hang

They call me hanging Johnny
hooray, hooray
but I never hung nobody
hang boys, hang

They say I must be crazy
hooray, hooray
the First Mate says maybe
hang boys, hang

Let's hang and haul together
hooray, hooray
and hang for better weather
hang boys, hang

Blow the Kaiser Down

Long drag chanty

Have you heard our King Georgy say
go blow the Kaiser down
we have heard our King Georgy say
go blow the Kaiser down

Go blow him down and batter his crown
go blow the Kaiser down
blow that German Kaiser down
go blow the Kaiser down

We'll sink his ships and bury them deep
go blow the Kaiser down
then this old world would peacefully sleep
go blow the Kaiser down

We have heard our King Georgy say
go blow the Kaiser down
he'll wish he never ruled the day
go blow the Kaiser down

We'll sink sink his bloody damn old ships
wherever they may be
and run that Kaiscr down aground
on the shores of Germany

He'll wish he never messed around
with us boy's, he'll see
the boys of the Royal Navy

The Dead Horse

Long drag chanty

Poor old man, your horse is going to die
and I say so, and I hope so;
Oh, poor old man,
your horse is going to die
Oh, poor—-old man

For ninety days I've ridden on him
and I say so, and I hope so;
when he dies I'll tan his skin
Oh, poor old man

If he lives we'll ride him again
And I say so, and I hope so;
We'll ride him again with a tighter rein
Oh, poor old man

It's up aloft the horse must go
and I say so, and I hope so;
We'll hoist him up, then bury him low,
Oh, poor old man.

One More Day

Pumpin' chanty

One more day
only one more day me Johnny
you can rock and roll me over
one more day

Can't you hear the old man growlin'
one more day
can't you hear the Mate a howlin'
one more day

Only one more day me Johnny
one more day
you can rock and roll me over
one more day

Can't you see the captain pacing

up and down the gangway
and the Bos'n he's a mumblin'
just one more day

Only one more day me Johnny
one more day
you can rock and roll me over
one more day

We've been sailin' for months now
haulin' and heavin' away
soon be sightin' port
just one more day

One More day (continued)

Only one more day me Johnny
one more day
you can rock and roll me over
one more day

I just can't a wait
to get me self ashore
'tis many a month
only one more day

Only one more day me Johnny
one more day
you can rock and roll me over
one more day

I've spent me life, oh Johnny
sailin' a far and away
a haulin' and a heavin'
for many a day

Only one more day me Johnny
one more day
you can rock and roll me over
one more day

Now we're headin' into port, me Johnny
heave away
just one more pull and then belay

Leave her, Johnny

Long drag chanty

Oh, the times are hard and the wages low
leave her Johnny, it's time to go
I'll pack my bags and go below
it's time for me, I just know

It's growl you may but go I must
leave her Johnny it's time to go
it matters not whether your first or last
it's time I've grown to old

I'm getting thin and growin' sad
leave her, there's nothing I can say
since first I joined this wooden clad
it's time, I can no longer stay

I thought I heard the First Mate say
leave her Johnny, leave her
one more drag and then belay
it's time, I can no longer stay

Rollin' Home

Old English Sea Song

Call all hands to man the capstan
see the cable run down clear
heave away and with a will boys
for old England we will steer

And we'll sing in joyful chorus
in the watches of the night
and we'll sight the shores of England
when the gray dawn meets the light

Rollin' home, rollin home,
rollin' home across the sea
rollin' home to dear old England,
rollin' home, dear land to thee

Up aloft amid the riggings
blows the loud exalting gale
like a birds wide out stretched pinions
spreads on high each swelling sail

And the wild waves cleft behind us
seem to murmur as they flow
there are lovin' hearts that wait you
in the land to which we go

Rolling Home (continued)

Rollin' home, rollin' home,
rollin' home across the sea
rollin' home to dear old England,
rollin' home, dear land to thee

Many thousand miles behind us
many thousand miles before
ancient oceans heave to waft us
to the well remembered shore

Rollin' home, rollin' home,
rollin' home across the sea
rollin' home to dear old,
rollin' home, dear land to thee

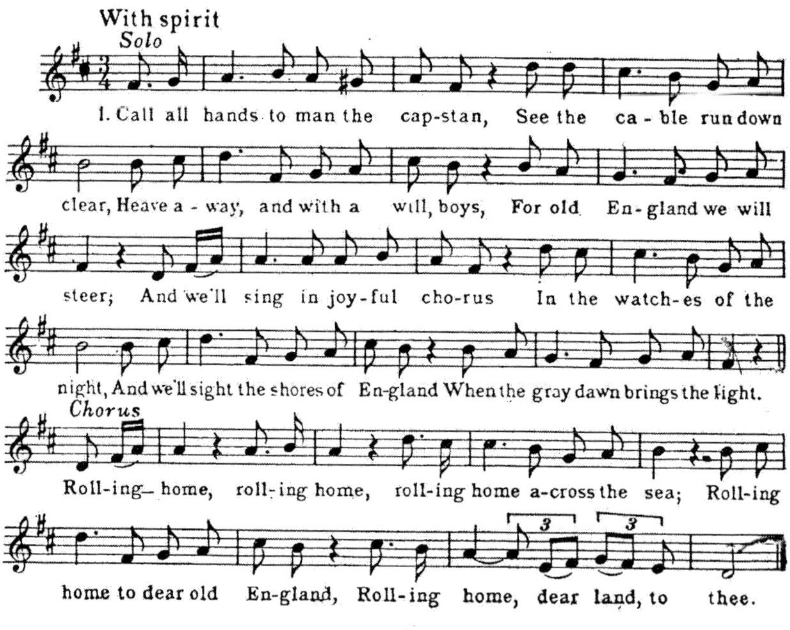

Storm Along

Pumpin' chanty

Stormy's gone, that good old man
to my way-a, storm along
old Stormy's gone, that good old man
to my way-a-a, Mister Storm along

We'll dig his grave with a silver spade
to my way-a, storm along
of the finest silk, his shroud be made
to my way-a-a, Mister Storm along

We'll lower him down with a golden chain
to my way-a, storm along
our eyes all dim but not with rain
to my way-a-a, Mister Storm along

He's moored at last and furled his sails
to my way-a, storm along
all free from wrecks and far from the gales
to my way-a-a, Mister Storm along

Old Stormy's heard the bugle call
to my way-a, storm along
to sing his dirge now one and all
to my way-a-a, Mister Storm along

Stormy's gone, that good old man
to my way-a, storm along
old Stormy's gone, that good old man
to my way-a-a, Mister Storm along

Glossary of Nautical Terms

able-bodied seaman	a sailor having some experience at sea, being of sound and strong body
aft	near or at the stern (rear) of a vessel
ahoy	to hail another, e.g. ship ahoy, ahoy Mate
aloft	in the top of the masts or on the higher yards, riggings
Aussie	abrev. for Australian
awl	a tool for making holes in fabric, leather, sails or repairing rigging
bark	a three or four masted ship, foremast and mainmast square rigged mizzenmast fore and aft rigged
bearing	a direction or heading of a ship
binnacle	a stand that contains the ships compass
block & tackle	a set of sheaves in a frame of wood provided with a hook used for lifting
boatswain	a ranking seaman in charge of rigging, anchors and cables
bow	the forward area of a ship
bowsprit	a large spar projecting from the bow
brig	a two masted square rigged ship
bucksaw	a saw in an H shaped frame used for sawing wood

cables	a strong braided rope used for fastening, mooring etc.
capstan	a drum turning on a spindle for moving heavy weights,
Captain	an officer in supreme charge of the conduct of the ship
chanty	a song sailors sing in rhythm to their work
chanty-man	a crew member of a ship familiar with the various chanties associated with different work tasks on board
clipper ship	a fast sleek sail ship with an overhanging bow and raking masts
coir	a stiff elastic fiber that comes from the outer crust of the coconut
cook stove	an iron stove, burning wood or coal used for cooking
crows nest	a platform high on a mast that serves as lookout station
current	a course or flow of waters, its onward motion
cutter	a broad stern boat with one mast and a jib and forestay sail
dinghy	a small light row boat carried on large ships like a life boat
'et'	slang word for ate
fathom	a unit of measure as found on a sounding line designating six foot intervals
First Mate	a deck officer in the Merchant Marine who ranks below Captain
forecastle	the forward part of a ship usually where the sailors have their bunks and stow gear
gaff boom	a spar on which the head of a fore and aft sail is extended

gale wind	a wind with velocity between 39 & 54 miles per hour
galley	the cooking area below deck where the crew's food is prepared
gangway	either side of an upper deck between the deck house and the rail and the quarter deck and forecastle
Gotland	an island in the Baltic Sea off the south east coast of Sweden
guy lines	a rope attached to a spar to hold steady
halyard	a rope or tackle used for hoisting sail
hard to port	a verbal command, meaning sharp turn of the rudder to the left
hawser	a large rope of three cables for towing or mooring
helmsman	a crew member assigned for steering the vessel
hemp	fiber from the hemp plant used for making rope
hitch to his trousers	a way of adjusting one's trousers, as a sudden movement or jerking, sailor style
hull	the body of a ship as to its particular characteristic
java	slang for coffee
jib	a triangular sail from the top of the foremast to the bowsprit or jib boom
ketch	a fore and aft rigged boat with mainmast and mizzenmast
knots	a unit of ships speed, as if a ship makes 8 nautical miles in an hour then the speed is estimated at 8 knots
lamptrimmer	a subordinate rank, usually assigned to lights and signals

mainsail	the principal sail on the mainmast in square rigged ships hanging from main yard
make fast	to tie up or secure to a mooring or wharf
manila rope	a fiber used to make rope from manila hemp
marline spike	a pointed iron tool used to splice rope
mate	a title descriptive of comradeship
mizzenmast	the aftermost mast in a two masted vessel and the third mast in a four or more masted vessel
mooring	a place where the ship is tied up or secured
mousings	a lashing of small stuff or spun yarn uniting the point and shank of a hook
North star	a northern hemisphere star towards which the axis of the earth is used used to determine positioning of a vessel
oar lock	a bracket mount that holds an oar for steering and rowing support
pilot	a qualified person who comes out and brings vessels in and out port
pipe up	to begin speaking, a Bos'ns whistle command used on board
port	larboard, the left side of a ship when facing the bow
pumps	bilge pumps below decks for pumping water from the holds
quarterdeck	part of the upper deck toward or at the stern

quartermaster	a petty officer who attends the helm, binnacle, ships signaling
ratlines	small three stranded rope attached to the shrouds forming a ladder
riggings	ropes that support spars and used to set and trim sails or standing rigging support masts
sail	a sheet of canvas or other durable fabric used to catch the wind
schooner	a fore and aft rigged ship, two masted with a smaller sail on the foremast and the mainmast stepped amidships
scuttlebutt	a naval term denoting rumor
sea legs	becoming accustomed to the rolling and pitching of a ship
seaman	a rank of sailor less experienced than able bodied seaman
sextant	an instrument that measures angular distances to observe altitude for finding longitude and latitude of ships position
shipshape	spotless, tidy, everything in its proper place
ships stations	various locations and duty assignments on board ship
shroud	a rope or pair of ropes leading from the masthead to give lateral support
shroud-laid	rope of four strands and laid right handed with a core of rope
sloop	a fore and aft rigged ship with one mast and single headsail jib
small stuff	name given to various small ropes and are designated by the number of strands and yarns used

sounding line	a line tied with knots at six foot intervals used for determining depth of a vessel in fathoms
spars	one of the main wooden parts of a mast usually fixed laterally to the mast from which a sail is hung
spanker	the fore and aft sail on the aftermast of square rigged ships
starboard	the right side of a vessel when facing the bow
steamer	a vessel propelled by steam
stops on sails	small lines that bind and secure a furled sail
thumpin' quid	a chaw of tobacco
topsail	in square rigged ships the sail next above the lower most sail, in fore and aft rigged ships, the sail set above
trawler	a fishing vessel
furled or unfurled	to role up or unrole a sail
wake	a track made by a ship as it sails thru the water
Wisby, Visby	a Swedish port on the island of Gotland
yard	a long spar tapered at the ends to support sail
zig-zagging	a nautical tactic to avoid being targeted

Gus's Sailing Logs

Swedish

British

Swedish Sailing Log

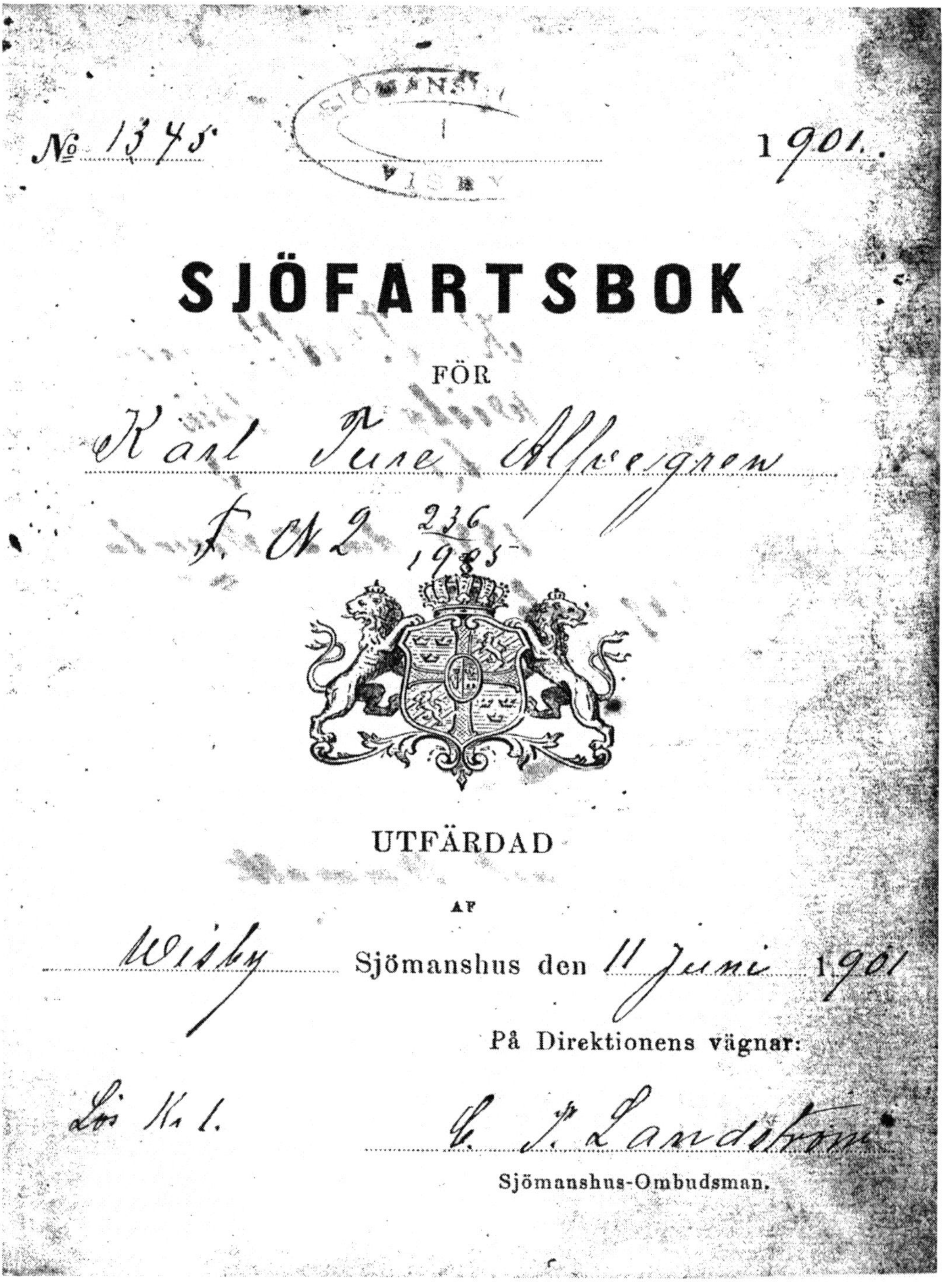

№ 1345 1901.

SJÖFARTSBOK

FÖR

Karl Ture Alfvegren

UTFÄRDAD

AF

Wisby Sjömanshus den 11 Juni 1901

På Direktionens vägnar:

C. J. Landström

Sjömanshus-Ombudsman.

Inskrifningsbok

(1902 års upplaga.)

för n:r F. 2236/1905

Karl Ture Alfregren

Född: d. 27 november 1884

Födelseort: Wisby

Yrke eller titel: Sjöman

Inskrifven d. 9/3 1905

Öfverförd

Inskrifven å sjömanshuset i: [illegible]

(införes med blyerts)

Boken utfärdad af:

[illegible]

Befälhafvare för Stockholms sjörullföringsområde.

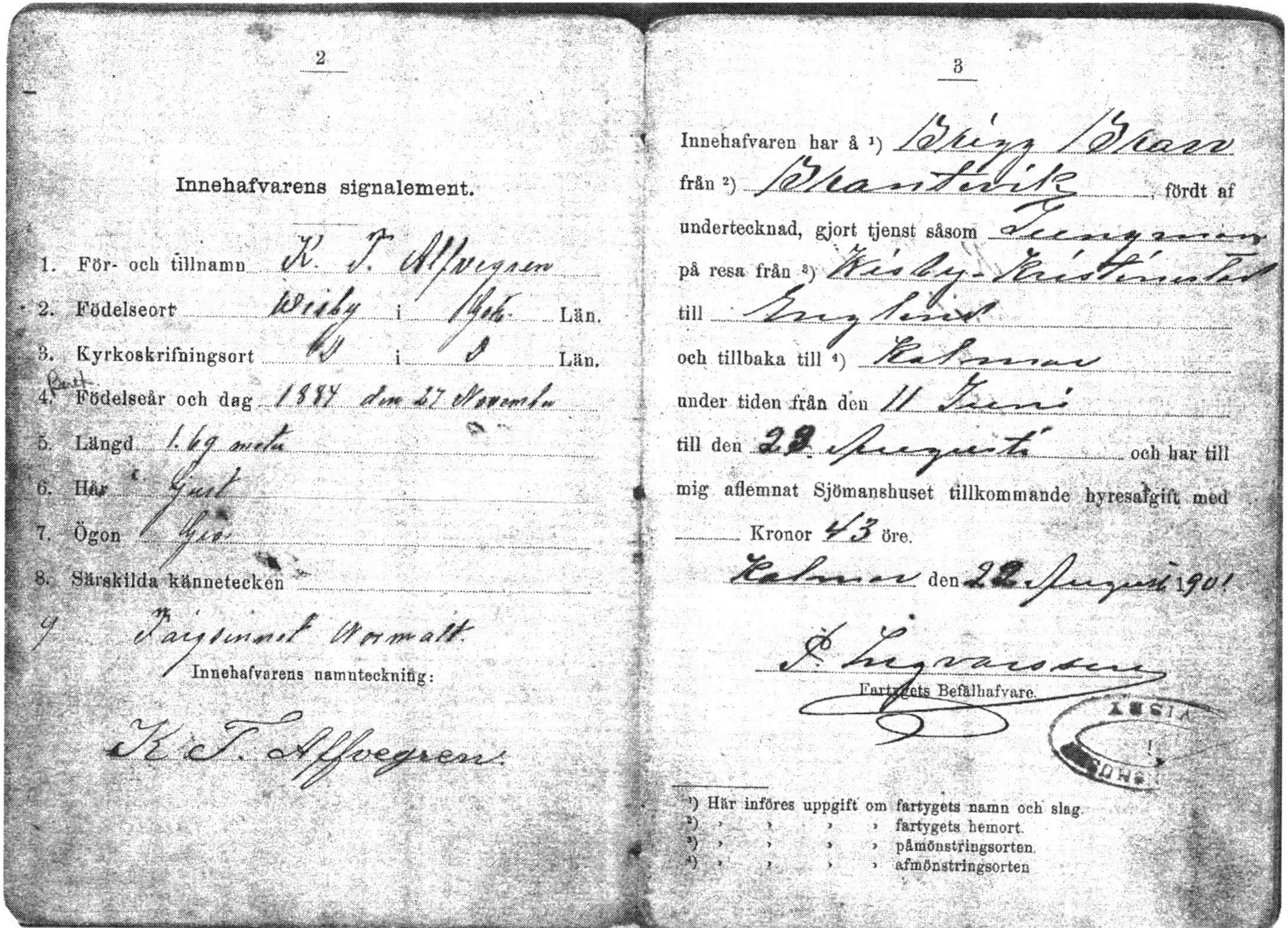

2

Innehafvarens signalement.

1. För- och tillnamn K. T. Alfvegren
2. Födelseort Wisby i ... Län.
3. Kyrkoskrifningsort ... i ... Län.
4. Födelseår och dag 1884 den 27 November
5. Längd 1.69 meter
6. Hår Gul
7. Ögon ...
8. Särskilda kännetecken ...

Innehafvarens namnteckning:

K. T. Alfvegren

3

Innehafvaren har å [1]) Brigg ...

från [2]) ..., fördt af

undertecknad, gjort tjenst såsom ...

på resa från [3]) Wisby ...

till ...

och tillbaka till [4]) Kalmar

under tiden från den 11 Juni

till den 23 Augusti och har till

mig aflemnat Sjömanshuset tillkommande hyresafgift med

... Kronor 43 öre.

Kalmar den 22 Augusti 1901.

...

Fartygets Befälhafvare.

[1]) Här införes uppgift om fartygets namn och slag.
[2]) » » » » fartygets hemort.
[3]) » » » » påmönstringsorten.
[4]) » » » » afmönstringsorten.

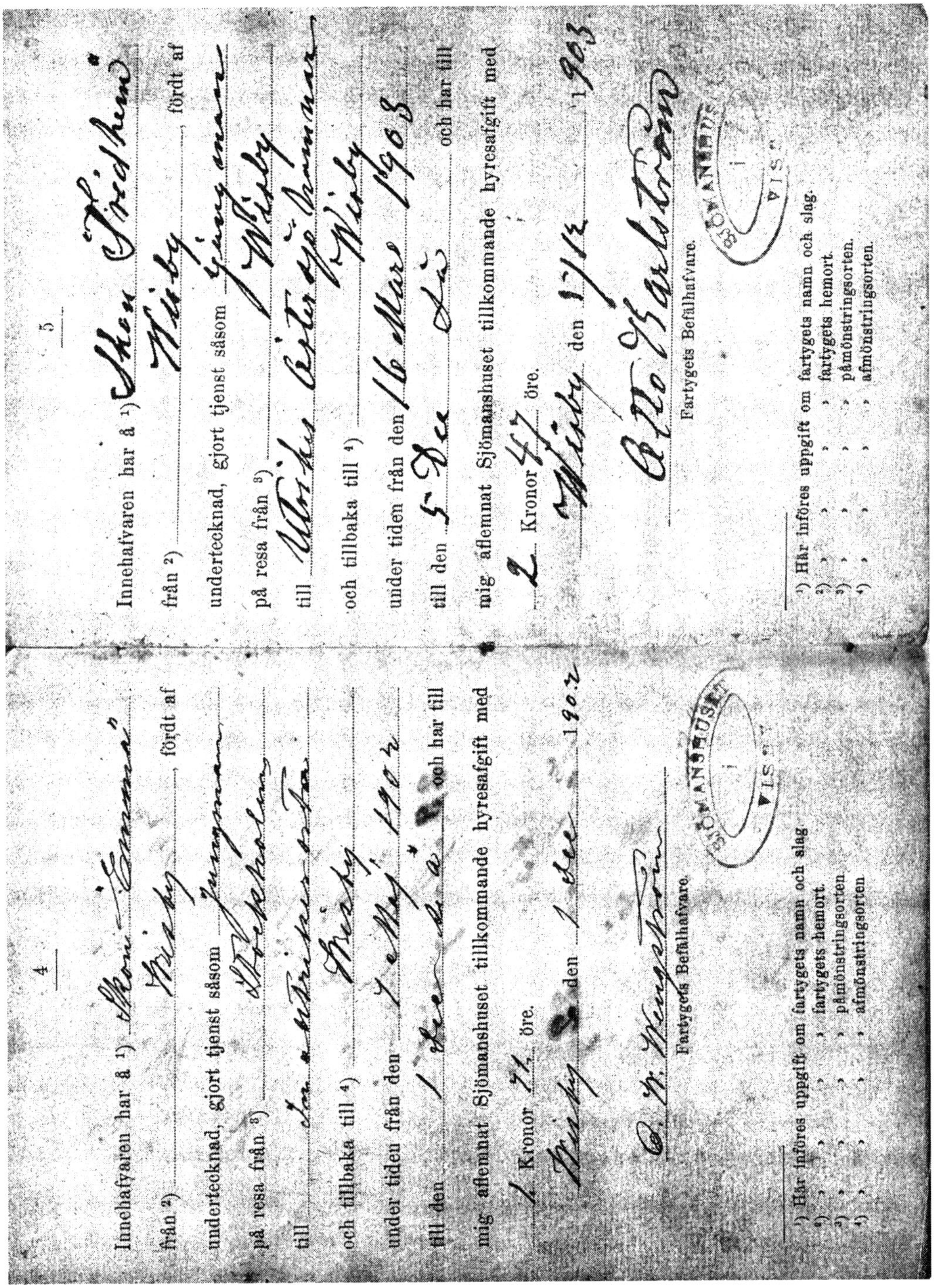

4

Innehafvaren har å [1]) Skon "Emma"

från [2]) Wisby, fördt af

undertecknad, gjort tjenst såsom Jungman

på resa från [3]) Stockholm

till [illegible]

och tillbaka till [4]) Wisby

under tiden från den 1 Maj 1902

till den 1 dec s. å. och har till

mig aflemnat Sjömanshuset tillkommande hyresafgift med

1 Kronor 71 öre.

Wisby den 1 dec 1902

C. W. Wengström

Fartygets Befälhafvare.

SJÖMANSHUSET i VISBY

[1]) Här införes uppgift om fartygets namn och slag.
[2]) » » » » fartygets hemort.
[3]) » » » » påmönstringsorten.
[4]) » » » » afmönstringsorten.

5

Innehafvaren har å [1]) Skon "Fridhem"

från [2]) Wisby, fördt af

undertecknad, gjort tjenst såsom Jungman

på resa från [3]) Wisby

till Utrikes Östersjöhamnar

och tillbaka till [4]) Wisby

under tiden från den 16 Mars 1903

till den 5 Dec s. å. och har till

mig aflemnat Sjömanshuset tillkommande hyresafgift med

2 Kronor 47 öre.

Wisby den 7/12 1903

Otto Karlström

Fartygets Befälhafvare.

SJÖMANSHUS i VISBY

[1]) Här införes uppgift om fartygets namn och slag.
[2]) » » » » fartygets hemort.
[3]) » » » » påmönstringsorten.
[4]) » » » » afmönstringsorten.

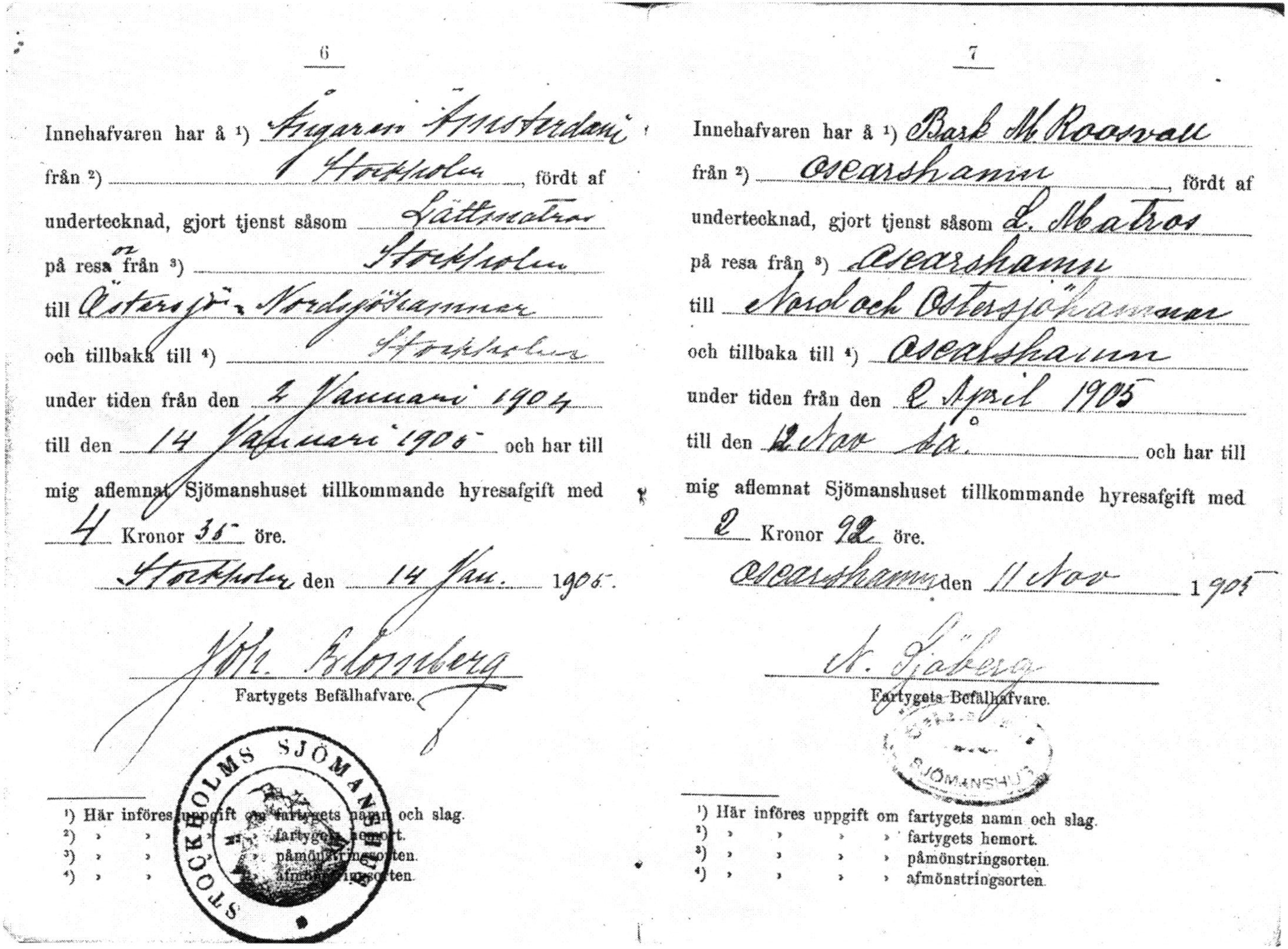

6

Innehafvaren har å [1]) Ångaren Amsterdam

från [2]) i Stockholm, fördt af

undertecknad, gjort tjenst såsom Lättmatros

på resor från [3]) Stockholm

till Östersjö- Nordsjöhamnar

och tillbaka till [4]) Stockholm

under tiden från den 2 Januari 1904

till den 14 Januari 1905 och har till

mig aflemnat Sjömanshuset tillkommande hyresafgift med

4 Kronor 35 öre.

Stockholm den 14 Jan. 1905.

Joh. Blomberg

Fartygets Befälhafvare.

Stockholms Sjömanshus

[1]) Här införes uppgift om fartygets namn och slag.
[2]) » » » » fartygets hemort.
[3]) » » » » påmönstringsorten.
[4]) » » » » afmönstringsorten.

7

Innehafvaren har å [1]) Bark M Roosvall

från [2]) Oscarshamn, fördt af

undertecknad, gjort tjenst såsom L. Matros

på resa från [3]) Oscarshamn

till Nord och Östersjöhamnar

och tillbaka till [4]) Oscarshamn

under tiden från den 2 April 1905

till den 12 Nov då och har till

mig aflemnat Sjömanshuset tillkommande hyresafgift med

2 Kronor 92 öre.

Oscarshamn den 11 Nov 1905

N. Sjöberg

Fartygets Befälhafvare.

Sjömanshus

[1]) Här införes uppgift om fartygets namn och slag.
[2]) » » » » fartygets hemort.
[3]) » » » » påmönstringsorten.
[4]) » » » » afmönstringsorten.

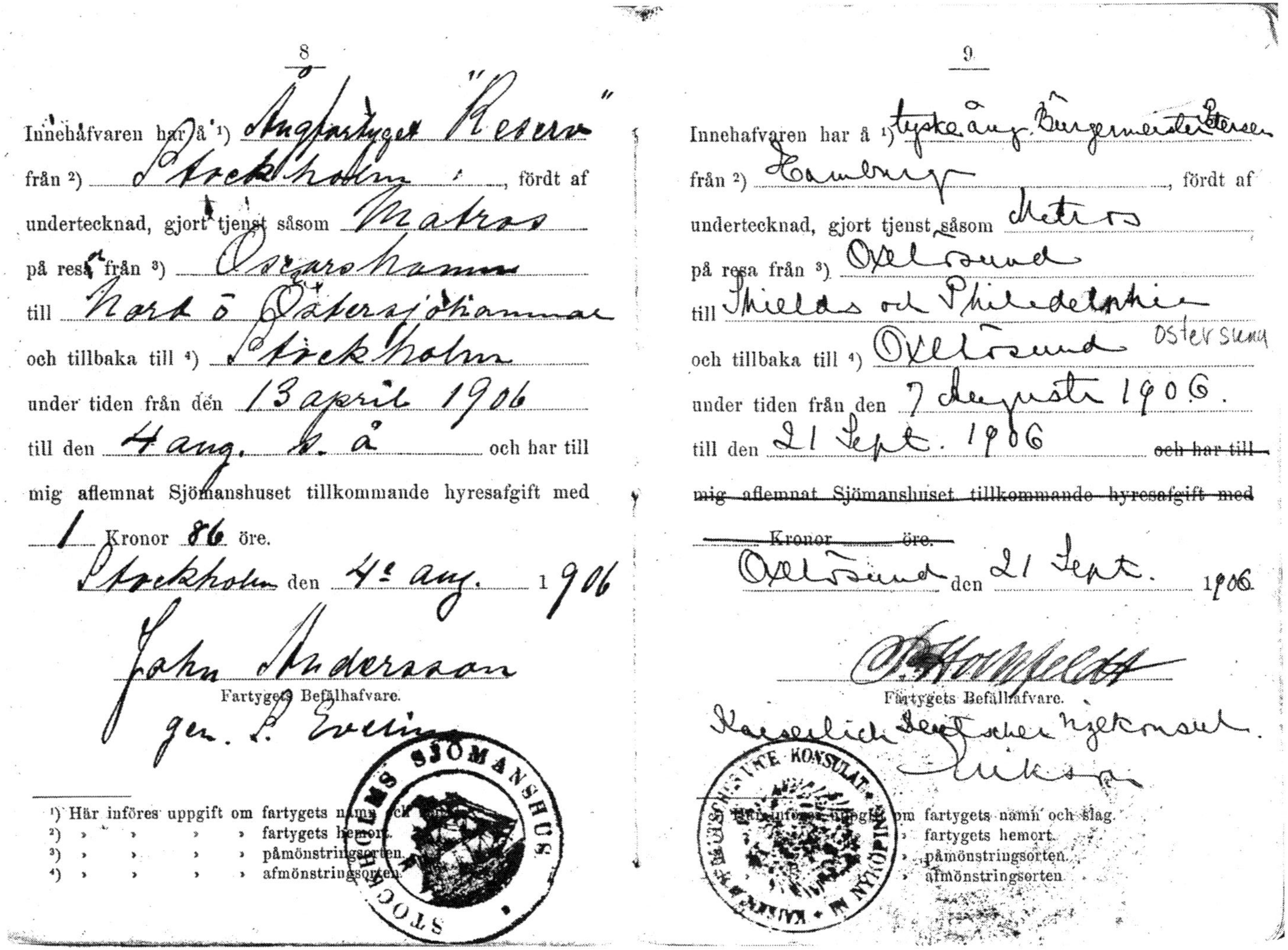

8

Innehåfvaren har å [1]) Ångfartyget "Reserv"

från [2]) Stockholm, fördt af

undertecknad, gjort tjenst såsom Matros

på resa från [3]) Oscarshamn

till Nord- o. Östersjöhamnar

och tillbaka till [4]) Stockholm

under tiden från den 13 april 1906

till den 4 aug. s. å och har till

mig aflemnat Sjömanshuset tillkommande hyresafgift med

1 Kronor 86 öre.

Stockholm den 4e aug. 1906

John Andersson

Fartygets Befälhafvare.

gen. P. Ekerlin

[1]) Här införes uppgift om fartygets namn och slag.
[2]) » » » » fartygets hemort.
[3]) » » » » påmönstringsorten.
[4]) » » » » afmönstringsorten.

9

Innehafvaren har å [1]) tyske ång. Burgemeister Petersen

från [2]) Hamburg, fördt af

undertecknad, gjort tjenst såsom Matros

på resa från [3]) Oxelösund

till Skields och Philadelphia

och tillbaka till [4]) Oxelösund Östersund

under tiden från den 7 Augusti 1906.

till den 21 Sept. 1906 ~~och har till~~

~~mig aflemnat Sjömanshuset tillkommande hyresafgift med~~

~~Kronor öre.~~

Oxelösund den 21 Sept. 1906

P. Hoffeldt

Fartygets Befälhafvare.

Kaiserlich Deutscher Vizekonsul.

Tiukson

[1]) Här införes uppgift om fartygets namn och slag.
[2]) » » » » fartygets hemort.
[3]) » » » » påmönstringsorten.
[4]) » » » » afmönstringsorten.

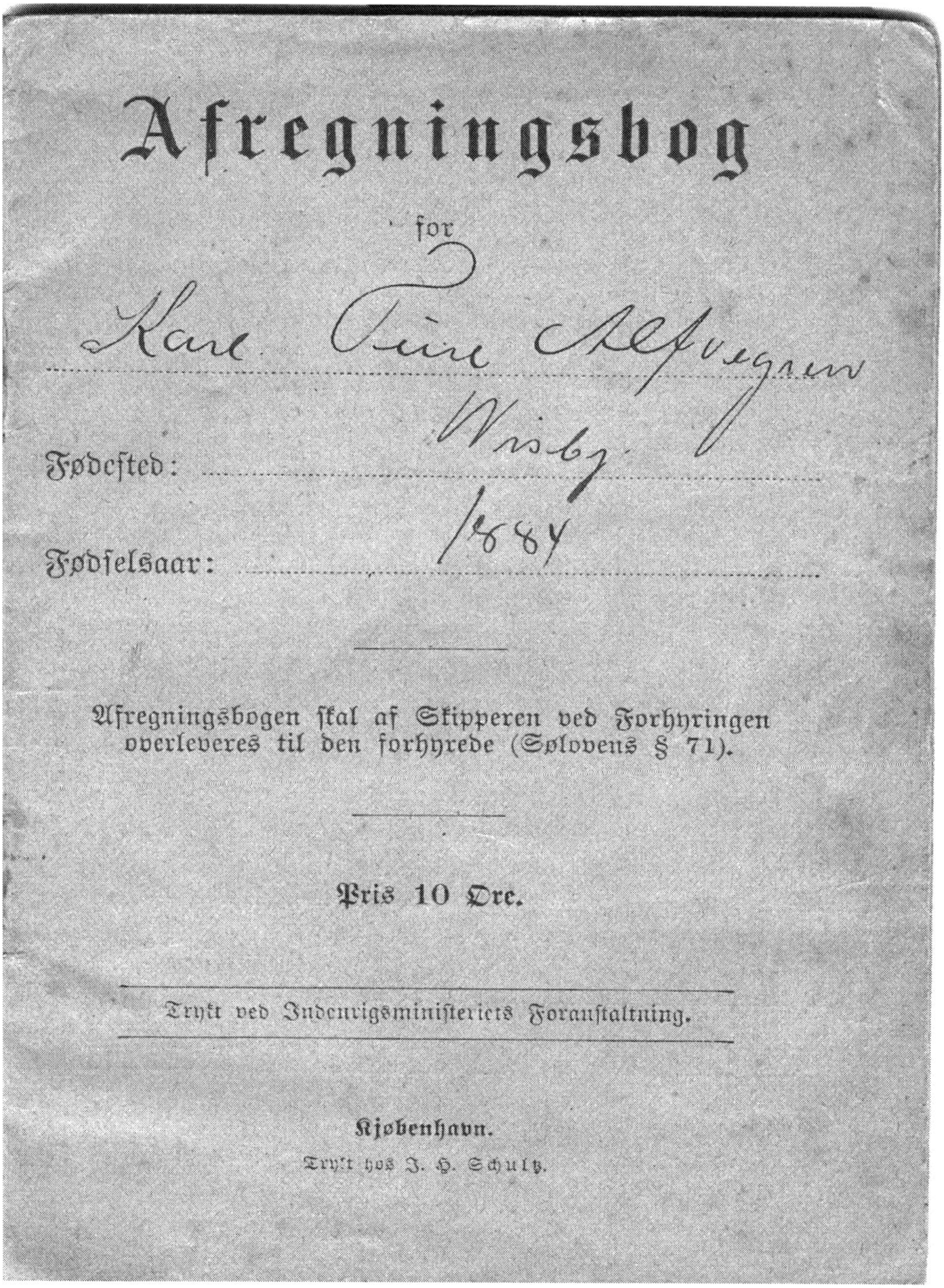

Afregningsbog

for

Karl Peter Alfvegren

Fødested: Wisby

Fødselsaar: 1884

Afregningsbogen skal af Skipperen ved Forhyringen overleveres til den forhyrede (Søloveus § 71).

Pris 10 Øre.

Trykt ved Indenrigsministeriets Foranstaltning.

Kjøbenhavn.

Trykt hos J. H. Schultz.

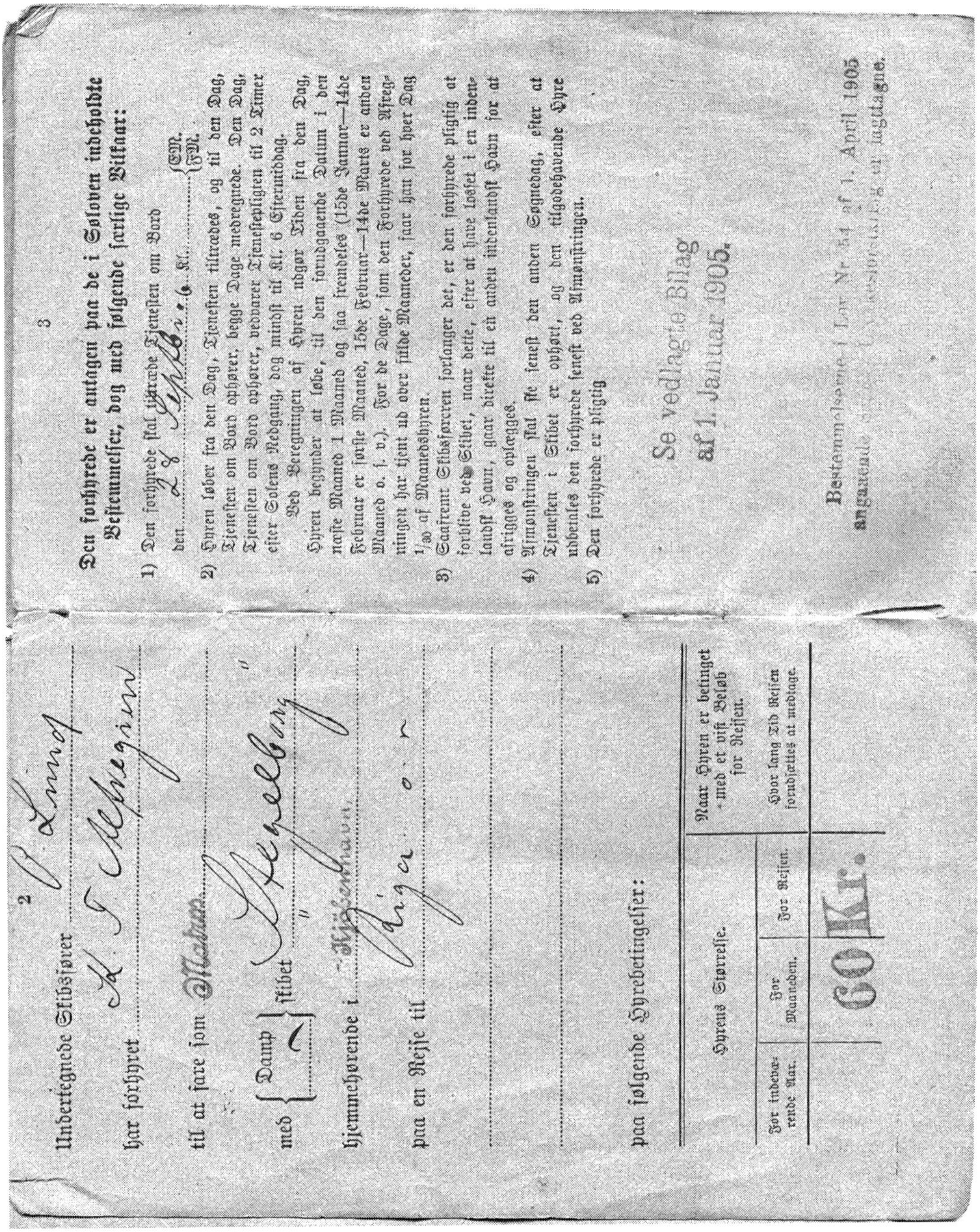

2

Undertegnede Skibsfører P. Lund

har forhyret K. T. Alfregren

til at fare som Matros

med {Damp / —} skibet "Stegelborg"

hjemmehørende i Kjøbenhavn

paa en Reise til Riga o. r

paa følgende Hyrebetingelser:

Hyrens Størrelse.			Naar Hyren er betinget med et vist Beløb for Reisen.
For indeværende Aar.	For Maaneden.	For Reisen.	Hvor lang Tid Reisen forudsættes at medtage.
	60 Kr.		

3

Den forhyrede er antagen paa de i Søloven indeholdte Bestemmelser, dog med følgende særlige Vilkaar:

1) Den forhyrede skal tiltræde Tjenesten om Bord den 28 Septbr. 6 Kl. {EM. / FM.}

2) Hyren løber fra den Dag, Tjenesten tiltrædes, og til den Dag, Tjenesten om Bord ophører, begge Dage medregnede. Den Dag, Tjenesten om Bord ophører, vedvarer Tjenestepligten til 2 Timer efter Solens Nedgang, dog mindst til Kl. 6 Eftermiddag.

Ved Beregningen af Hyren udgør Tiden fra den Dag, Hyren begynder at løbe, til den forudgaaende Datum i den næste Maaned 1 Maaned og saa fremdeles (15de Januar—14de Februar er første Maaned, 15de Februar—14de Marts er anden Maaned o. s. v.). For de Dage, som den Forhyrede ved Afregningen har tjent ud over fulde Maaneder, faar han for hver Dag $^1/_{30}$ af Maanedshyren.

3) Saafremt Skibsføreren forlanger det, er den forhyrede pligtig at forblive ved Skibet, naar dette, efter at have losset i en indenlandsk Havn, gaar direkte til en anden indenlandsk Havn for at afrigges og oplægges.

4) Afmønstringen skal ske senest den anden Søgnedag, efter at Tjenesten i Skibet er ophørt, og den tilgodehavende Hyre udbetales den forhyrede senest ved Afmønstringen.

5) Den forhyrede er pligtig

Se vedlagte Bilag
af 1. Januar 1905.

Bestemmelserne i Lov Nr. 54 af 1. April 1905 angaaende [illegible] Ulykkesforsikring er iagttagne.

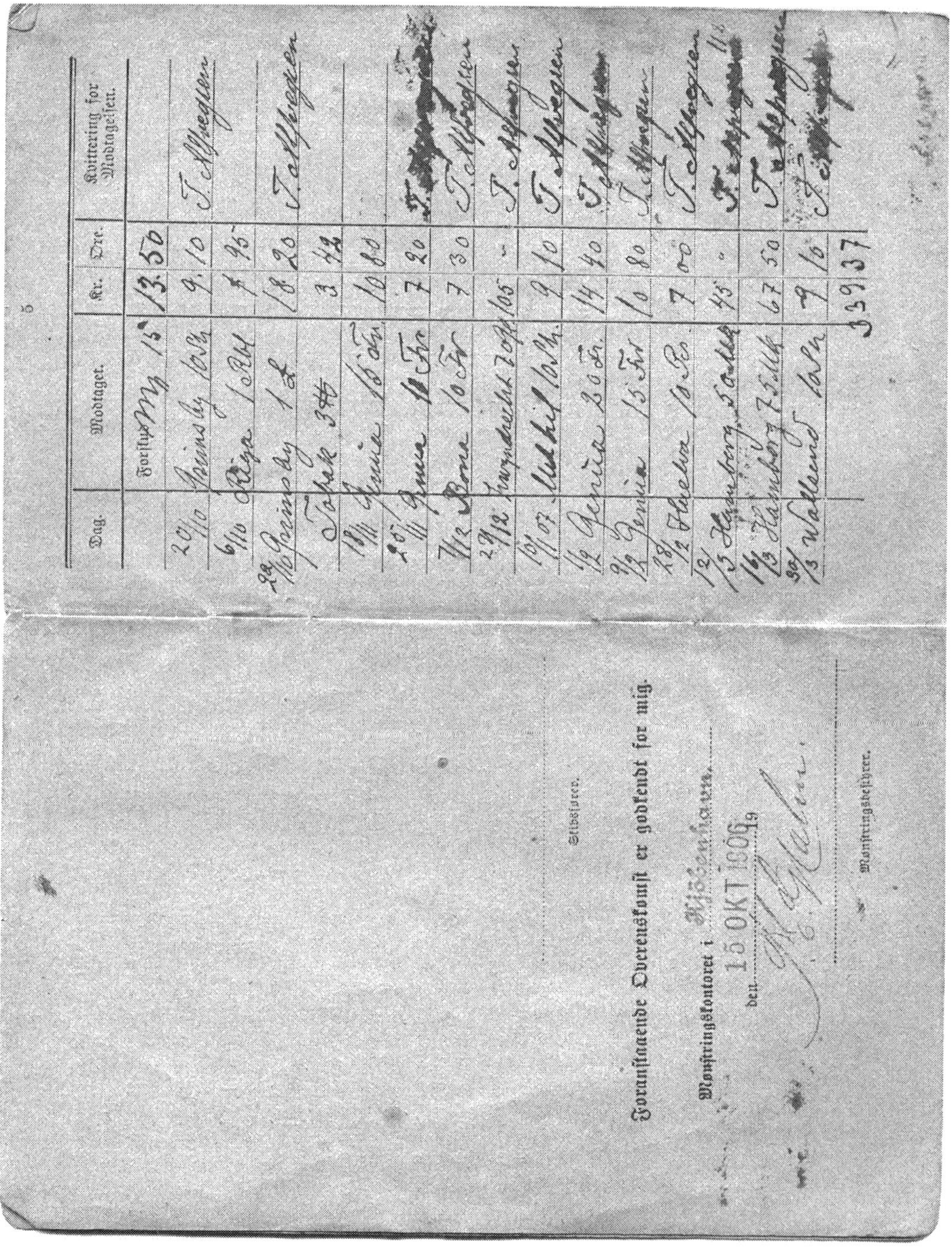

Skibsfører.

Foranstaaende Overenskomst er godkendt for mig.

Mønstringskontoret i Kjøbenhavn

den 15 OKT 1906 19

Mønstringsbestyrer.

5

Dag.	Modtaget.	Kr.	Øre.	Kvittering for Modtagelsen.
	Forskud 15	13	50	
20/10	Grimsby 10 Sh.	9	10	T. Alfsvegen
6/10	Riga 1 Rbl	[illegible]	95	
29/10	Grimsby 1 £	18	20	T. Alfsvegen
	Tobak 3 ℔	3	42	
18/11	Genua 15 Fr.	10	80	
25/11	Genua 10 Fr.	7	20	T. Alfsvegen
7/12	Bona 10 Fr.	7	30	T. Alfsvegen
29/12	Frugdrecht 7 off	105	–	T. Alfsvegen
10/1 07	Mobil 10 Sh.	9	10	T. Alfsvegen
1/2	Genua 20 Fr.	14	40	T. Alfsvegen
9/2	Genua 15 Fr.	10	80	T. Alfsvegen
28/2	Huelva 10 Ps.	7	00	T. Alfsvegen
12/3	Hamburg 50 Mk	45	–	T. Alfsvegen
16/3	Hamburg 75 Mk	67	50	T. Alfsvegen
30/3	Wallsend 10 Sh.	9	10	T. Alfsvegen
		339	37	

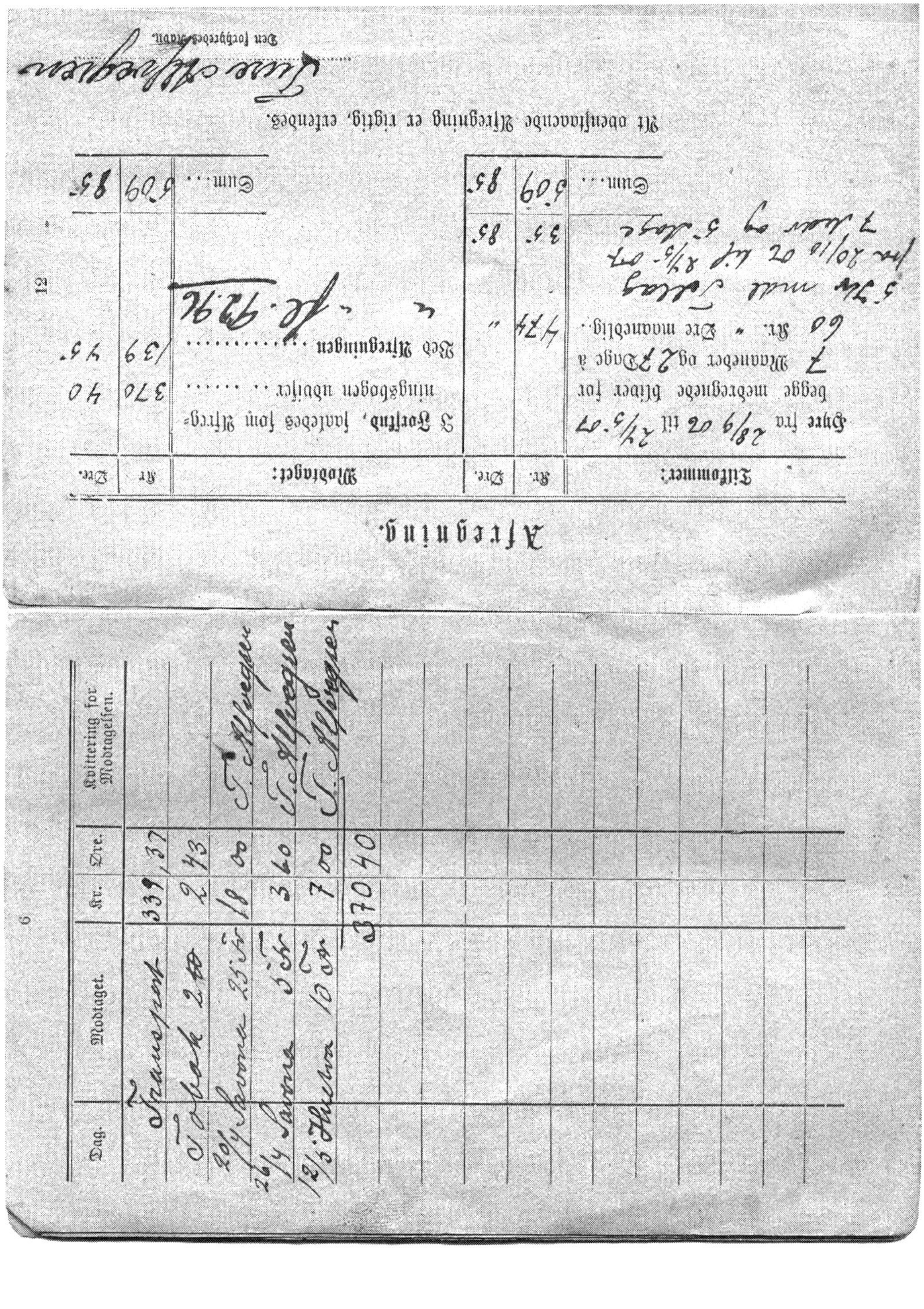

12

Afregning.

Tilkommer:	Kr.	Øre.	Modtaget:	Kr.	Øre.
Hyre fra 28/9 02 til 27/5 07 begge medregnede bliver for 7 Maaneder og 27 Dage à 60 Kr. " Øre maanedlig..	474	"	I Forskud, saaledes som Afregningsbogen udviser	370	40
5 Tdr mel Tillæg fra 20/10 02 til 27/5 07 7 Maaneder og 5 Dage	35	85	Ved Afregningen	139	45
			K.R.N.		
Sum...	509	85	Sum...	509	85

At ovenstaaende Afregning er rigtig, erkendes.

[signature] Alfregnen

Den forhyredes Navn.

6

Dag.	Modtaget.	Kr.	Øre.	Kvittering for Modtagelsen.
	Transport	339	37	
	Tobak 2.40	2	43	
20/4	Lanna 25 kr	18	00	T. Alfregnen
26/4	Lanna 5 kr	3	60	T. Alfregnen
18/5	Hlema 10 kr	7	00	T. Alfregnen
		370	40	

British

Board

Of Trade

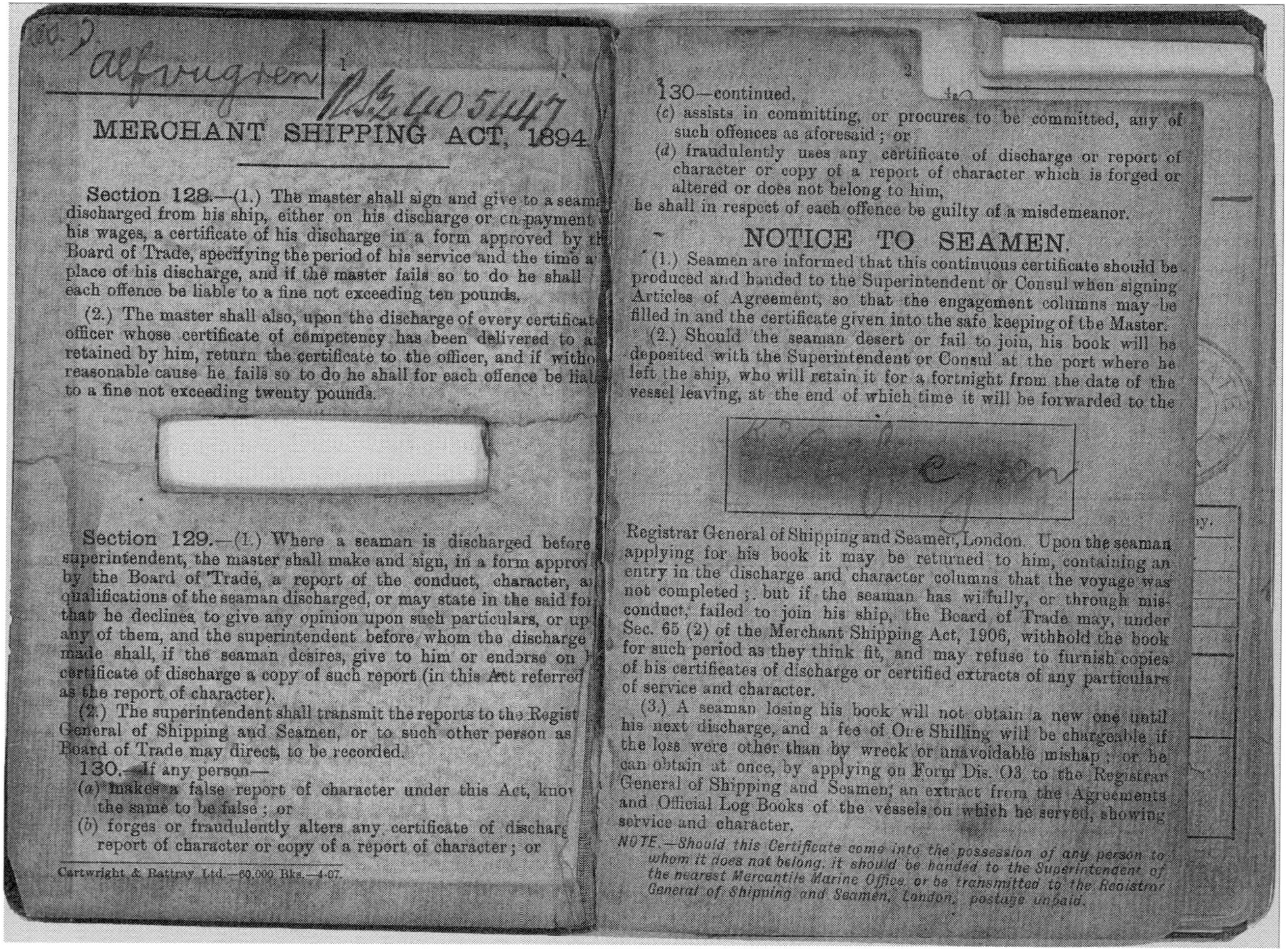

1

MERCHANT SHIPPING ACT, 1894.

Section 128.—(1.) The master shall sign and give to a seam[illegible] discharged from his ship, either on his discharge or on payment [illegible] his wages, a certificate of his discharge in a form approved by th[illegible] Board of Trade, specifying the period of his service and the time a[illegible] place of his discharge, and if the master fails so to do he shall [illegible] each offence be liable to a fine not exceeding ten pounds.

(2.) The master shall also, upon the discharge of every certificat[illegible] officer whose certificate of competency has been delivered to a[illegible] retained by him, return the certificate to the officer, and if witho[illegible] reasonable cause he fails so to do he shall for each offence be liab[illegible] to a fine not exceeding twenty pounds.

Section 129.—(1.) Where a seaman is discharged before [illegible] superintendent, the master shall make and sign, in a form approv[illegible] by the Board of Trade, a report of the conduct, character, a[illegible] qualifications of the seaman discharged, or may state in the said for[illegible] that he declines to give any opinion upon such particulars, or up[illegible] any of them, and the superintendent before whom the discharge [illegible] made shall, if the seaman desires, give to him or endorse on [illegible] certificate of discharge a copy of such report (in this Act referred [illegible] as the report of character).

(2.) The superintendent shall transmit the reports to the Regist[illegible] General of Shipping and Seamen, or to such other person as [illegible] Board of Trade may direct, to be recorded.

130.—If any person—

(*a*) makes a false report of character under this Act, kno[illegible] the same to be false; or

(*b*) forges or fraudulently alters any certificate of discharg[illegible] report of character or copy of a report of character; or

Cartwright & Rattray Ltd.—60,000 Bks.—4-07.

2

130—continued.

(*c*) assists in committing, or procures to be committed, any of such offences as aforesaid; or

(*d*) fraudulently uses any certificate of discharge or report of character or copy of a report of character which is forged or altered or does not belong to him,

he shall in respect of each offence be guilty of a misdemeanor.

NOTICE TO SEAMEN.

(1.) Seamen are informed that this continuous certificate should be produced and handed to the Superintendent or Consul when signing Articles of Agreement, so that the engagement columns may be filled in and the certificate given into the safe keeping of the Master.

(2.) Should the seaman desert or fail to join, his book will be deposited with the Superintendent or Consul at the port where he left the ship, who will retain it for a fortnight from the date of the vessel leaving, at the end of which time it will be forwarded to the Registrar General of Shipping and Seamen, London. Upon the seaman applying for his book it may be returned to him, containing an entry in the discharge and character columns that the voyage was not completed; but if the seaman has wilfully, or through misconduct, failed to join his ship, the Board of Trade may, under Sec. 65 (2) of the Merchant Shipping Act, 1906, withhold the book for such period as they think fit, and may refuse to furnish copies of his certificates of discharge or certified extracts of any particulars of service and character.

(3.) A seaman losing his book will not obtain a new one until his next discharge, and a fee of One Shilling will be chargeable if the loss were other than by wreck or unavoidable mishap; or he can obtain at once, by applying on Form Dis. O3 to the Registrar General of Shipping and Seamen, an extract from the Agreements and Official Log Books of the vessels on which he served, showing service and character.

NOTE.—Should this Certificate come into the possession of any person to whom it does not belong, it should be handed to the Superintendent of the nearest Mercantile Marine Office, or be transmitted to the Registrar General of Shipping and Seamen, London, postage unpaid.

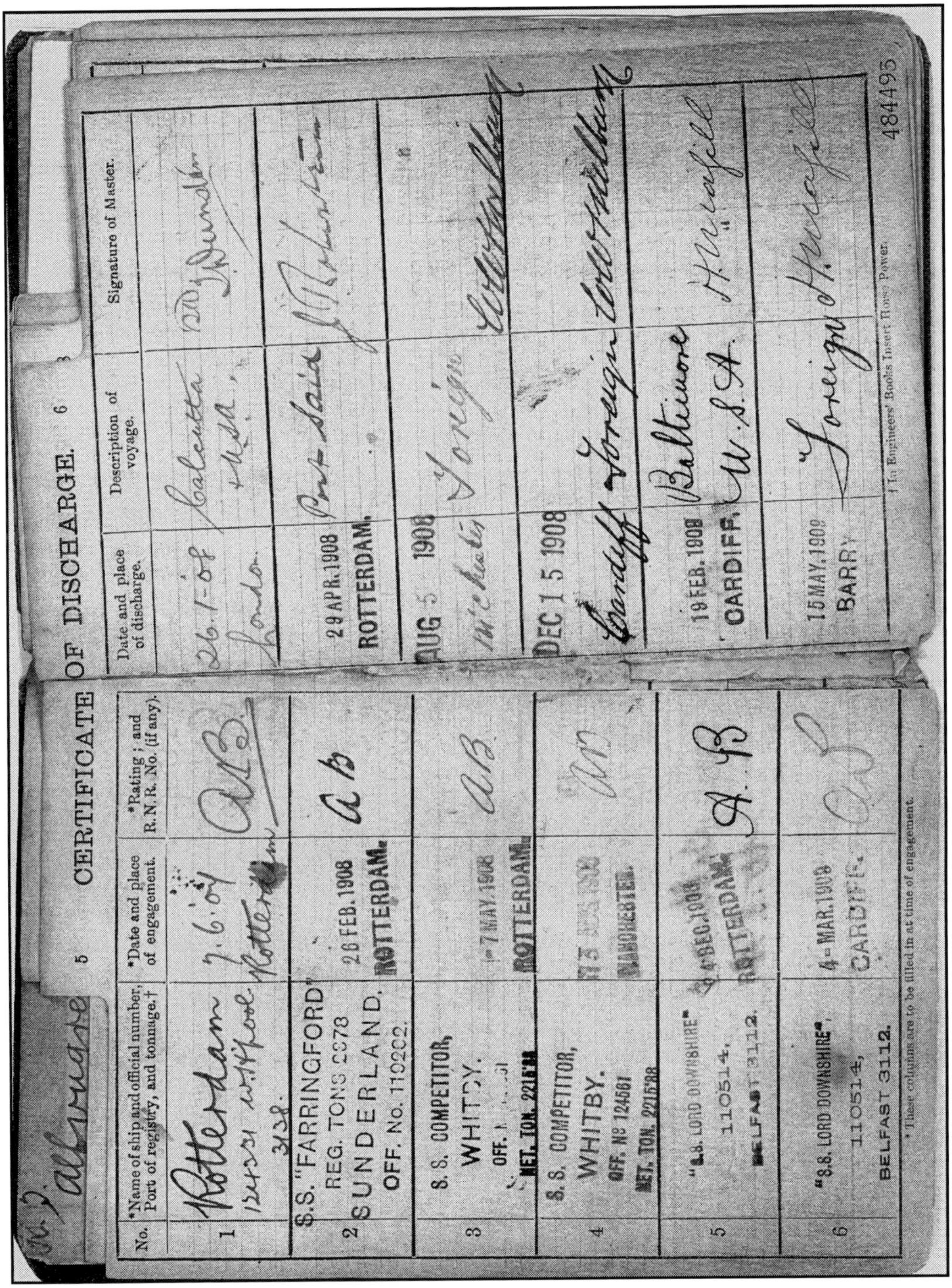

5 CERTIFICATE OF DISCHARGE. 6

No.	*Name of ship and official number, Port of registry, and tonnage.†	*Date and place of engagement.	*Rating; and R.N.R. No. (if any).	Date and place of discharge.	Description of voyage.	Signature of Master.
1	Rotterdam 124531 Liverpool 3138.	7.6.07 Rotterdam	AB	26.1.08 London	Calcutta India	[illegible]
2	S.S. "FARRINGFORD" REG. TONS 2378 SUNDERLAND. OFF. No. 119202	26 FEB. 1908 ROTTERDAM.	AB	29 APR. 1908 ROTTERDAM.	Port Said	[illegible]
3	S.S. COMPETITOR, WHITBY. OFF. [illegible] NET. TON. [illegible]	7 MAY. 1908 ROTTERDAM.	AB	AUG 5 1908 [illegible]	Foreign	[illegible]
4	S.S. COMPETITOR, WHITBY. OFF. No 124561 NET. TON. 2215.38	[illegible] MANCHESTER.	AB	DEC 15 1908 Cardiff	Foreign	[illegible]
5	"S.S. LORD DOWNSHIRE" 110514. BELFAST 3112.	4 DEC. 1908 ROTTERDAM.	A.B	19 FEB. 1908 CARDIFF.	Baltimore U.S.A.	[illegible]
6	"S.S. LORD DOWNSHIRE", 110514, BELFAST 3112.	4 MAR. 1909 CARDIFF.	AB	15 MAY. 1909 BARRY	Foreign	[illegible]

* These columns are to be filled in at time of engagement.

† In Engineers' Books Insert Horse Power.

484495

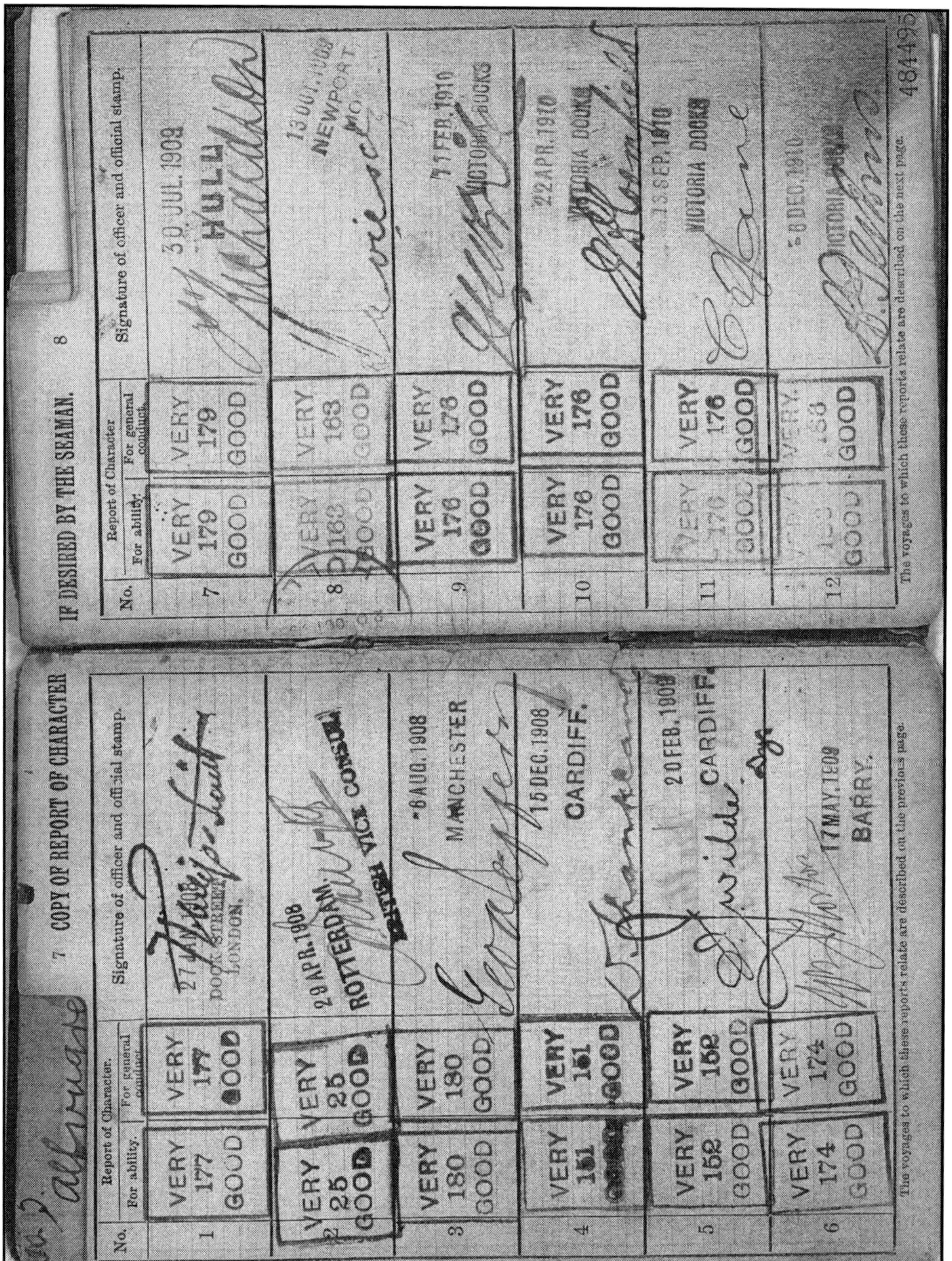

7 COPY OF REPORT OF CHARACTER

No.	Report of Character. For ability.	For general conduct.	Signature of officer and official stamp.
1	VERY 177 GOOD	VERY 177 GOOD	27 JAN 1908 DOCK STREET LONDON.
2	VERY 25 GOOD	VERY 25 GOOD	29 APR. 1908 ROTTERDAM. BRITISH VICE CONSUL
3	VERY 180 GOOD	VERY 180 GOOD	8 AUG. 1908 MANCHESTER
4	VERY 151 [illegible]	VERY 151 GOOD	15 DEC. 1908 CARDIFF.
5	VERY 152 GOOD	VERY 152 GOOD	20 FEB. 1909 CARDIFF.
6	VERY 174 GOOD	VERY 174 GOOD	17 MAY 1909 BARRY.

The voyages to which these reports relate are described on the previous page.

IF DESIRED BY THE SEAMAN. 8

No.	Report of Character For ability.	For general conduct.	Signature of officer and official stamp.
7	VERY 179 GOOD	VERY 179 GOOD	30 JUL. 1909 HULL
8	VERY 163 GOOD	VERY 163 GOOD	13 OCT. 1909 NEWPORT MON.
9	VERY 176 GOOD	VERY 178 GOOD	11 FEB. 1910 VICTORIA DOCKS
10	VERY 176 GOOD	VERY 176 GOOD	22 APR. 1910 VICTORIA DOCKS
11	VERY 176 GOOD	VERY 176 GOOD	13 SEP. 1910 VICTORIA DOCKS
12	VERY 153 GOOD	VERY 153 GOOD	8 DEC. 1910 VICTORIA DOCKS

The voyages to which these reports relate are described on the next page.

484495

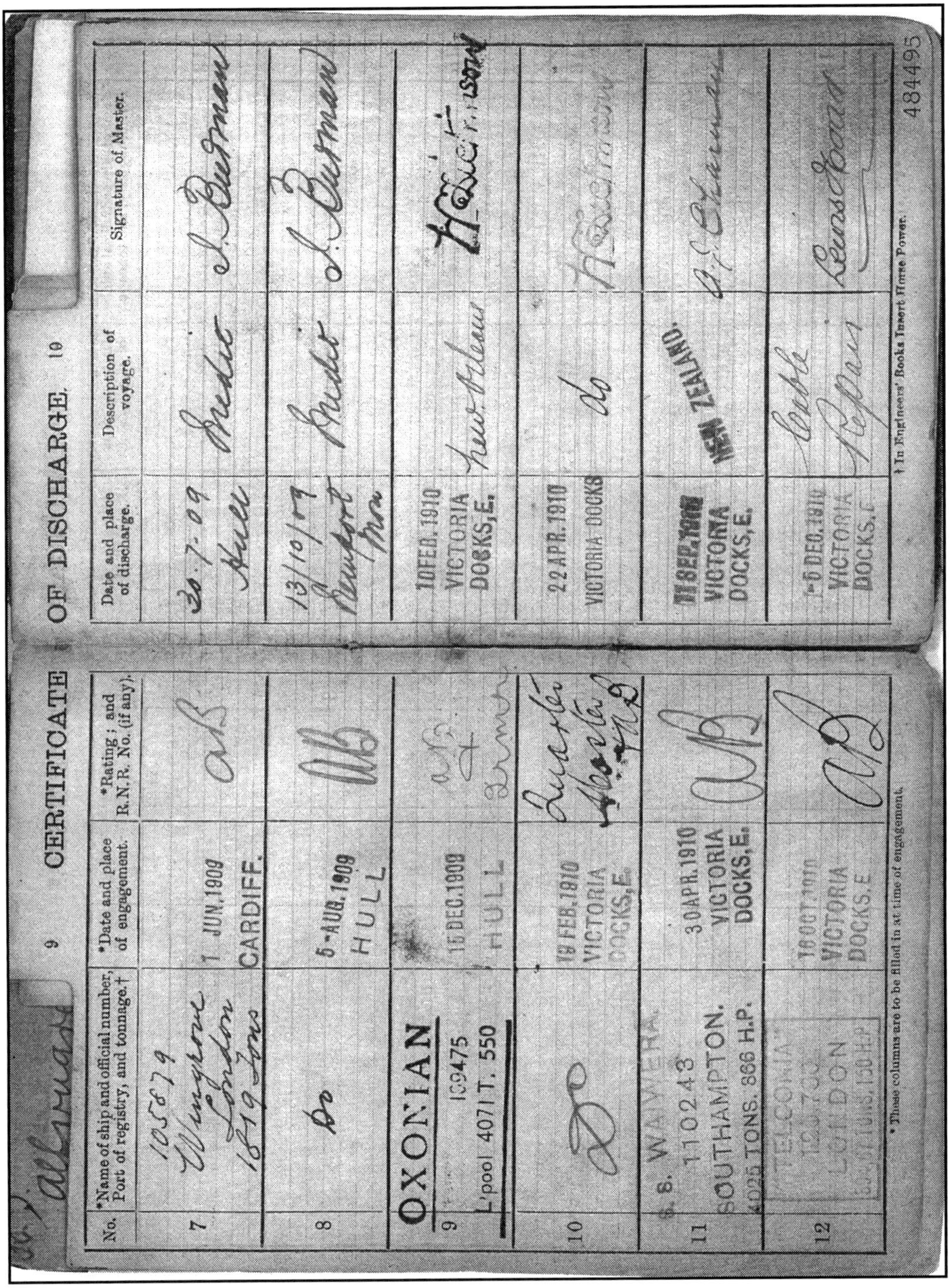

9 CERTIFICATE OF DISCHARGE 10

No.	*Name of ship and official number, Port of registry, and tonnage.†	*Date and place of engagement.	*Rating; and R.N.R. No. (if any).	Date and place of discharge.	Description of voyage.	Signature of Master.
7	105679 Wingrove London 1819 Tons	1 JUN.1909 CARDIFF.	AB	20-7-09 Hull	[illegible]	J. Burman
8	Do	5-AUG.1909 HULL	AB	13/10/09 Newport Mon	[illegible]	J. Burman
9	OXONIAN 109475 L'pool 4071 T. 550	15 DEC.1909 HULL	[illegible]	10 FEB.1910 VICTORIA DOCKS,E.	New Orleans	[illegible]
10	Do	19 FEB.1910 VICTORIA DOCKS,E.	Quarter Master	22 APR.1910 VICTORIA DOCKS	do	[illegible]
11	S.S. WAIMERA 110243 SOUTHAMPTON. 4025 TONS. 866 H.P.	30 APR.1910 VICTORIA DOCKS,E.	AB	11 SEP.1910 VICTORIA DOCKS,E.	NEW ZEALAND.	[illegible]
12	TELCONIA [illegible] LONDON [illegible] TONS. 150 H.P.	18 OCT.1910 VICTORIA DOCKS,E	AB	5 DEC.1910 VICTORIA DOCKS,E	Cable Repairs	[illegible]

* These columns are to be filled in at time of engagement.

† In Engineers' Books Insert Horse Power.

484495

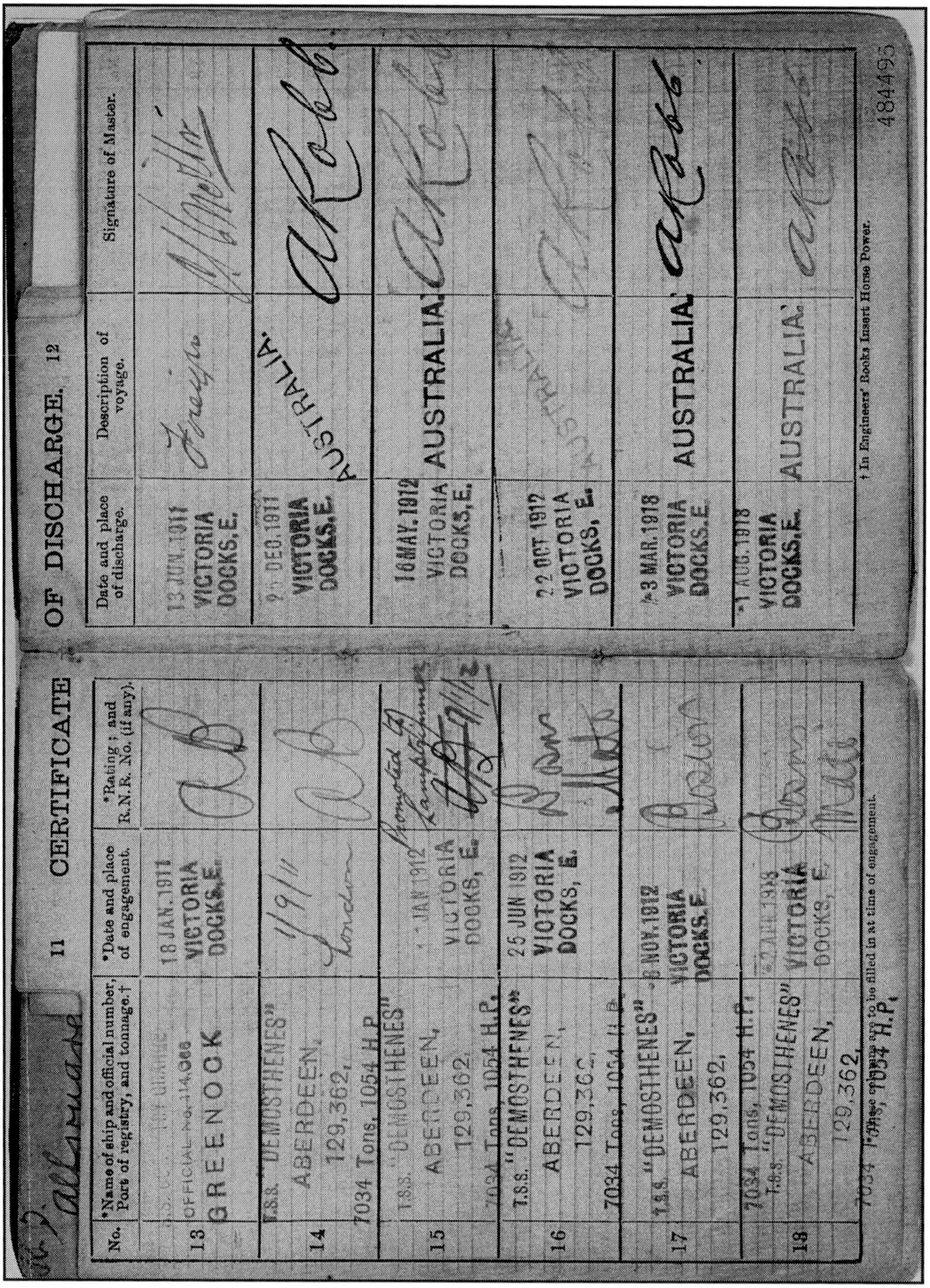

11 CERTIFICATE OF DISCHARGE. 12

No.	*Name of ship and official number, Port of registry, and tonnage.†	*Date and place of engagement.	*Rating; and R.N.R. No. (if any).	Date and place of discharge.	Description of voyage.	Signature of Master.
13	OFFICIAL No. 114,066 GREENOCK	18 JAN. 1911 VICTORIA DOCKS, E.	AB	13 JUN. 1911 VICTORIA DOCKS, E.	Foreign	[illegible]
14	T.S.S. "DEMOSTHENES" ABERDEEN, 129,362, 7034 Tons, 1054 H.P.	1/9/11 London	AB	9 DEC. 1911 VICTORIA DOCKS, E.	AUSTRALIA.	A Robb
15	T.S.S. "DEMOSTHENES" ABERDEEN, 129,362, 7034 Tons, 1054 H.P.	JAN 1912 VICTORIA DOCKS, E.	Promoted to Lamptrimmer AB 17/1/12	16 MAY. 1912 VICTORIA DOCKS, E.	AUSTRALIA.	A Robb
16	T.S.S. "DEMOSTHENES" ABERDEEN, 129,362, 7034 Tons, 1054 H.P.	25 JUN 1912 VICTORIA DOCKS, E.	Bosns Mate	22 OCT 1912 VICTORIA DOCKS, E.	AUSTRALIA.	A Robb
17	T.S.S. "DEMOSTHENES" ABERDEEN, 129,362, 7034 Tons, 1054 H.P.	8 NOV. 1912 VICTORIA DOCKS, E.	Bosun	3 MAR. 1913 VICTORIA DOCKS, E.	AUSTRALIA.	A Robb
18	T.S.S. "DEMOSTHENES" ABERDEEN, 129,362, 7034 Tons, 1054 H.P.	APR 1913 VICTORIA DOCKS, E.	Bosns Mate	1 AUG. 1913 VICTORIA DOCKS, E.	AUSTRALIA.	A Robb

* These columns are to be filled in at time of engagement.

† In Engineers' Books Insert Horse Power.

484495

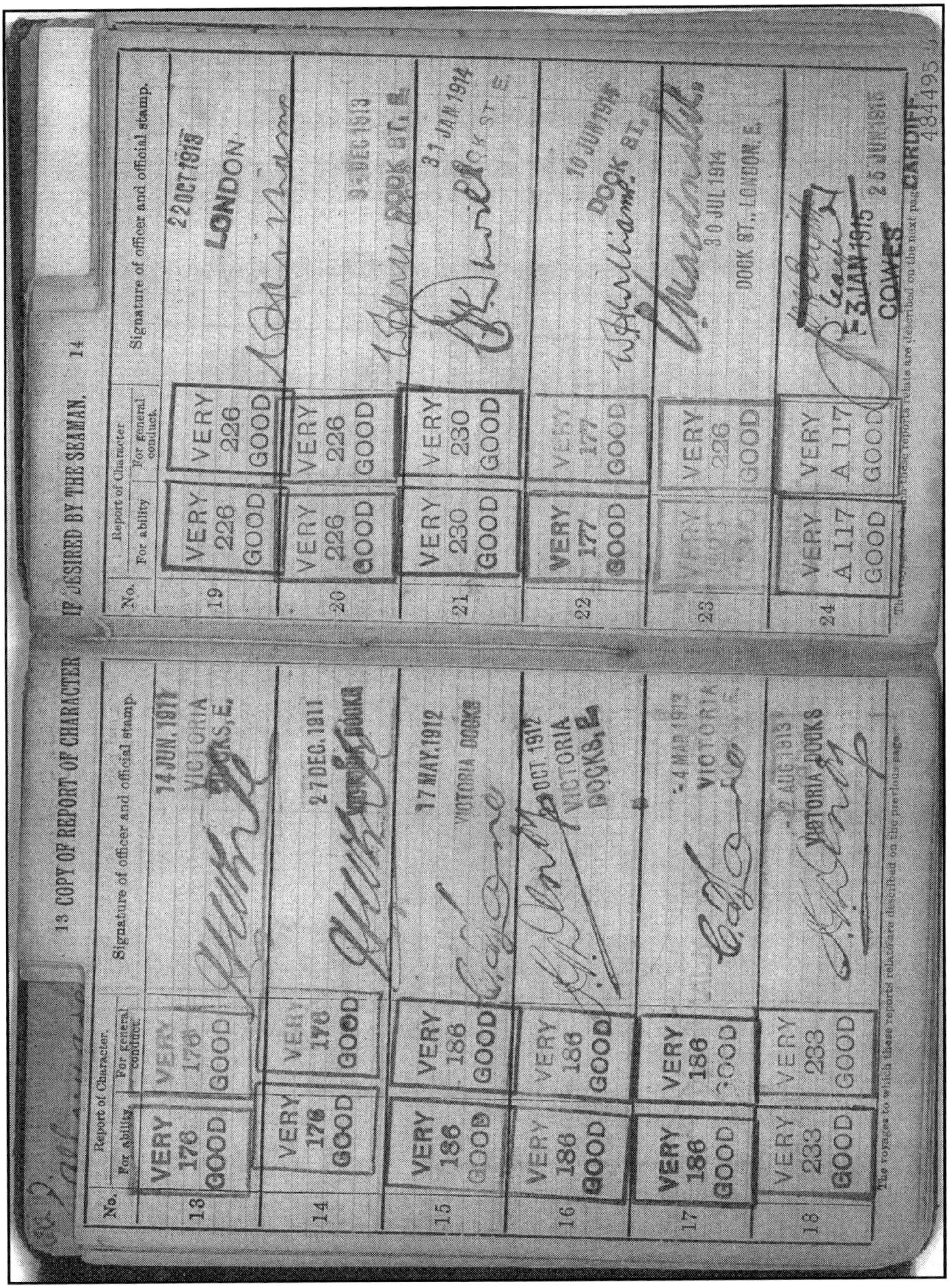

13 COPY OF REPORT OF CHARACTER IF DESIRED BY THE SEAMAN. 14

No.	Report of Character. For ability	For general conduct.	Signature of officer and official stamp.
13	VERY 176 GOOD	VERY 176 GOOD	14 JUN. 1911 VICTORIA DOCKS, E.
14	VERY 176 GOOD	VERY 176 GOOD	27 DEC. 1911 VICTORIA DOCKS
15	VERY 186 GOOD	VERY 186 GOOD	17 MAY. 1912 VICTORIA DOCKS
16	VERY 186 GOOD	VERY 186 GOOD	23 OCT. 1912 VICTORIA DOCKS, E.
17	VERY 186 GOOD	VERY 186 GOOD	4 MAR 1913 VICTORIA DOCKS, E.
18	VERY 233 GOOD	VERY 233 GOOD	2 AUG 1913 VICTORIA DOCKS

The voyages to which these reports relate are described on the previous page.

No.	Report of Character For ability	For general conduct.	Signature of officer and official stamp.
19	VERY 226 GOOD	VERY 226 GOOD	22 OCT 1913 LONDON
20	VERY 226 GOOD	VERY 226 GOOD	9 DEC 1913 DOCK ST. E.
21	VERY 230 GOOD	VERY 230 GOOD	31 JAN 1914 DOCK ST. E.
22	VERY 177 GOOD	VERY 177 GOOD	W.H. Williams. 10 JUN 1914 DOCK ST. E.
23	[illegible]	VERY 226 GOOD	30 JUL 1914 DOCK ST., LONDON, E.
24	VERY A 117 GOOD	VERY A 117 GOOD	3 JAN 1915 COWES 25 JUN 1915 CARDIFF.

The voyages to which these reports relate are described on the next page.

484495

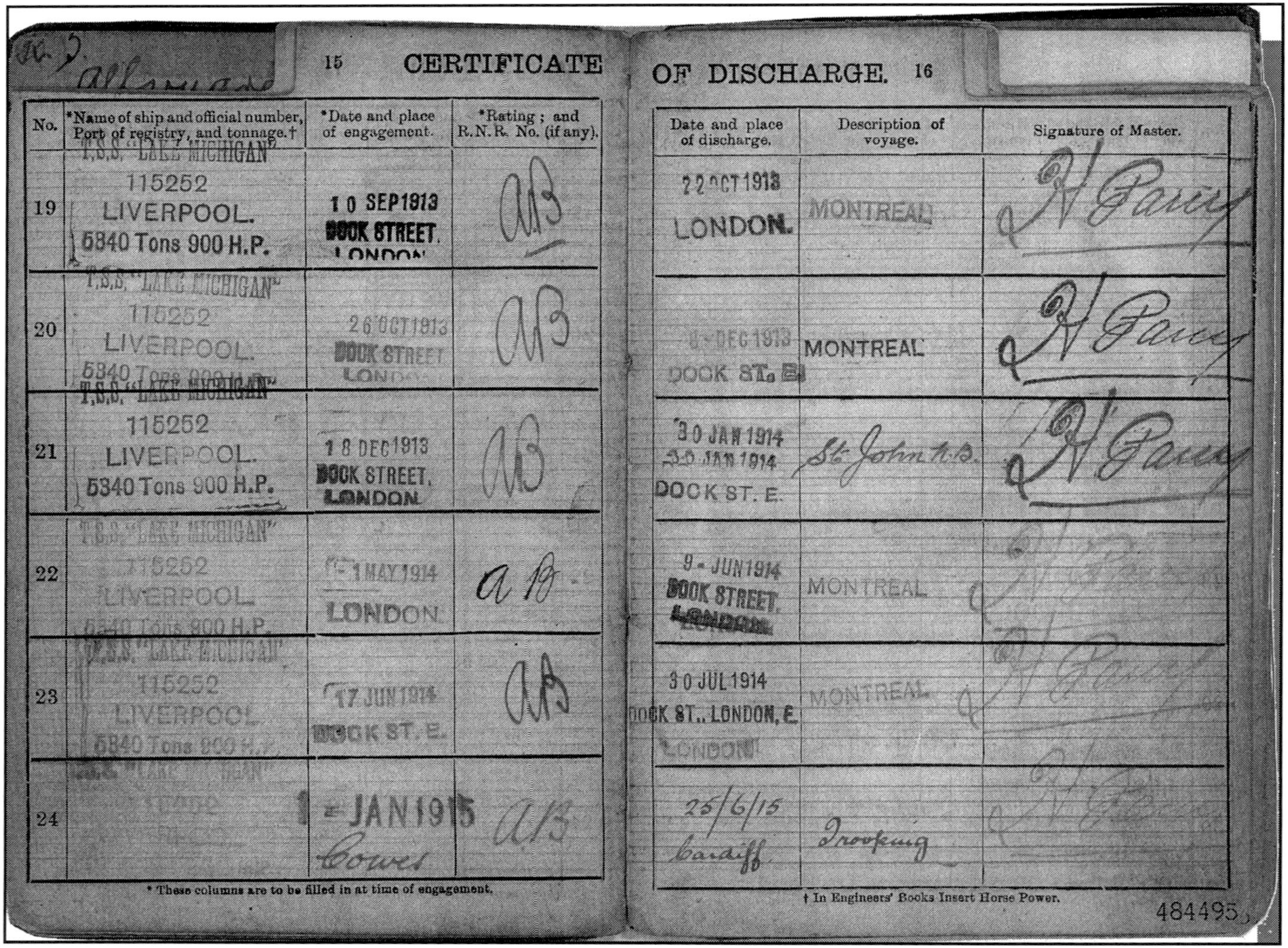

15 **CERTIFICATE OF DISCHARGE.** 16

No.	*Name of ship and official number, Port of registry, and tonnage.†	*Date and place of engagement.	*Rating; and R.N.R. No. (if any).	Date and place of discharge.	Description of voyage.	Signature of Master.
19	T.S.S. "LAKE MICHIGAN" 115252 LIVERPOOL. 5340 Tons 900 H.P.	10 SEP 1913 DOCK STREET. LONDON	AB	22 OCT 1913 LONDON.	MONTREAL	H. Pacey
20	T.S.S. "LAKE MICHIGAN" 115252 LIVERPOOL. 5340 Tons 900 H.P.	26 OCT 1913 DOCK STREET LONDON	AB	[illegible] DEC 1913 DOCK ST. E.	MONTREAL	H. Pacey
21	T.S.S. "LAKE MICHIGAN" 115252 LIVERPOOL. 5340 Tons 900 H.P.	18 DEC 1913 DOCK STREET. LONDON	AB	30 JAN 1914 30 JAN 1914 DOCK ST. E.	St. John N.B.	H. Pacey
22	T.S.S. "LAKE MICHIGAN" 115252 LIVERPOOL. 5340 Tons 900 H.P.	[illegible] MAY 1914 LONDON	AB	9 JUN 1914 DOCK STREET. LONDON	MONTREAL	
23	T.S.S. "LAKE MICHIGAN" 115252 LIVERPOOL. 5340 Tons 900 H.P.	17 JUN 1914 DOCK ST. E.	AB	30 JUL 1914 DOCK ST., LONDON, E. LONDON	MONTREAL	
24	[illegible]	1 JAN 1915 Cowes	AB	25/6/15 Cardiff	Trooping	

* These columns are to be filled in at time of engagement.

† In Engineers' Books Insert Horse Power.

484495

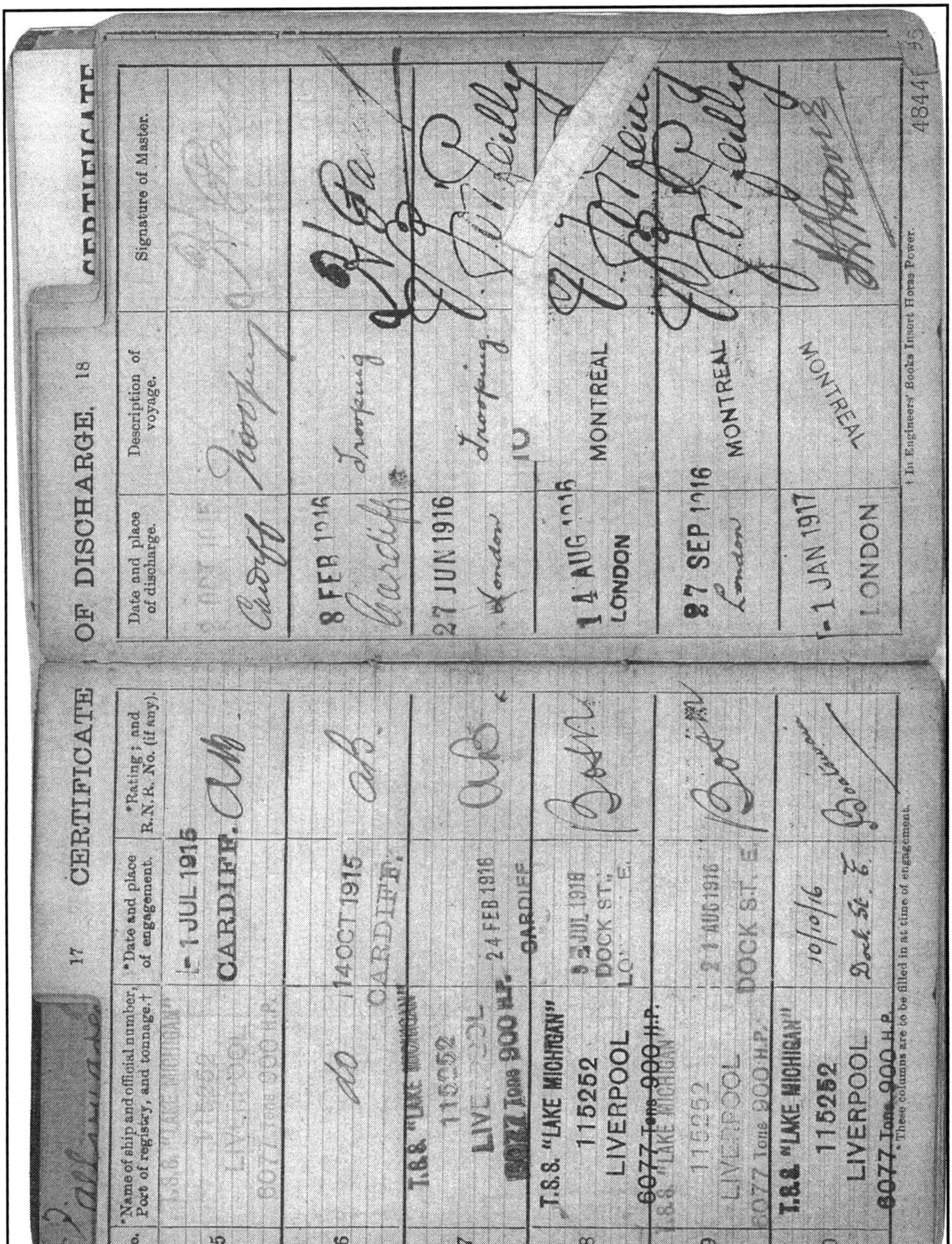

17 CERTIFICATE OF DISCHARGE. 18

No.	*Name of ship and official number, Port of registry, and tonnage.†	*Date and place of engagement.	*Rating; and R.N.R. No. (if any).	Date and place of discharge.	Description of voyage.	Signature of Master.
25	T.S.S. "LAKE MICHIGAN" 115252 LIVERPOOL 6077 Tons 900 H.P.	1 JUL 1915 CARDIFF	AB	[illegible] Cardiff	Trooping	[illegible]
26	do	14 OCT 1915 CARDIFF	ab	8 FEB 1916 Cardiff	Trooping	[illegible]
27	T.S.S. "LAKE MICHIGAN" 115252 LIVERPOOL 6077 Tons 900 H.P.	24 FEB 1916 CARDIFF	AB	27 JUN 1916 London	Trooping	[illegible]
28	T.S.S. "LAKE MICHIGAN" 115252 LIVERPOOL 6077 Tons 900 H.P.	3 JUL 1916 DOCK ST. LONDON E.	Bosn	1 AUG 1916 LONDON	MONTREAL	[illegible]
29	T.S.S. "LAKE MICHIGAN" 115252 LIVERPOOL 6077 Tons 900 H.P.	21 AUG 1916 DOCK ST. E.	Bosn	27 SEP 1916 London	MONTREAL	[illegible]
30	T.S.S. "LAKE MICHIGAN" 115252 LIVERPOOL 6077 Tons 900 H.P.	10/10/16 Dock St. E.	Boatswain	1 JAN 1917 LONDON	MONTREAL	[illegible]

*These columns are to be filled in at time of engagement.

† In Engineers' Books insert Horse Power.

4844

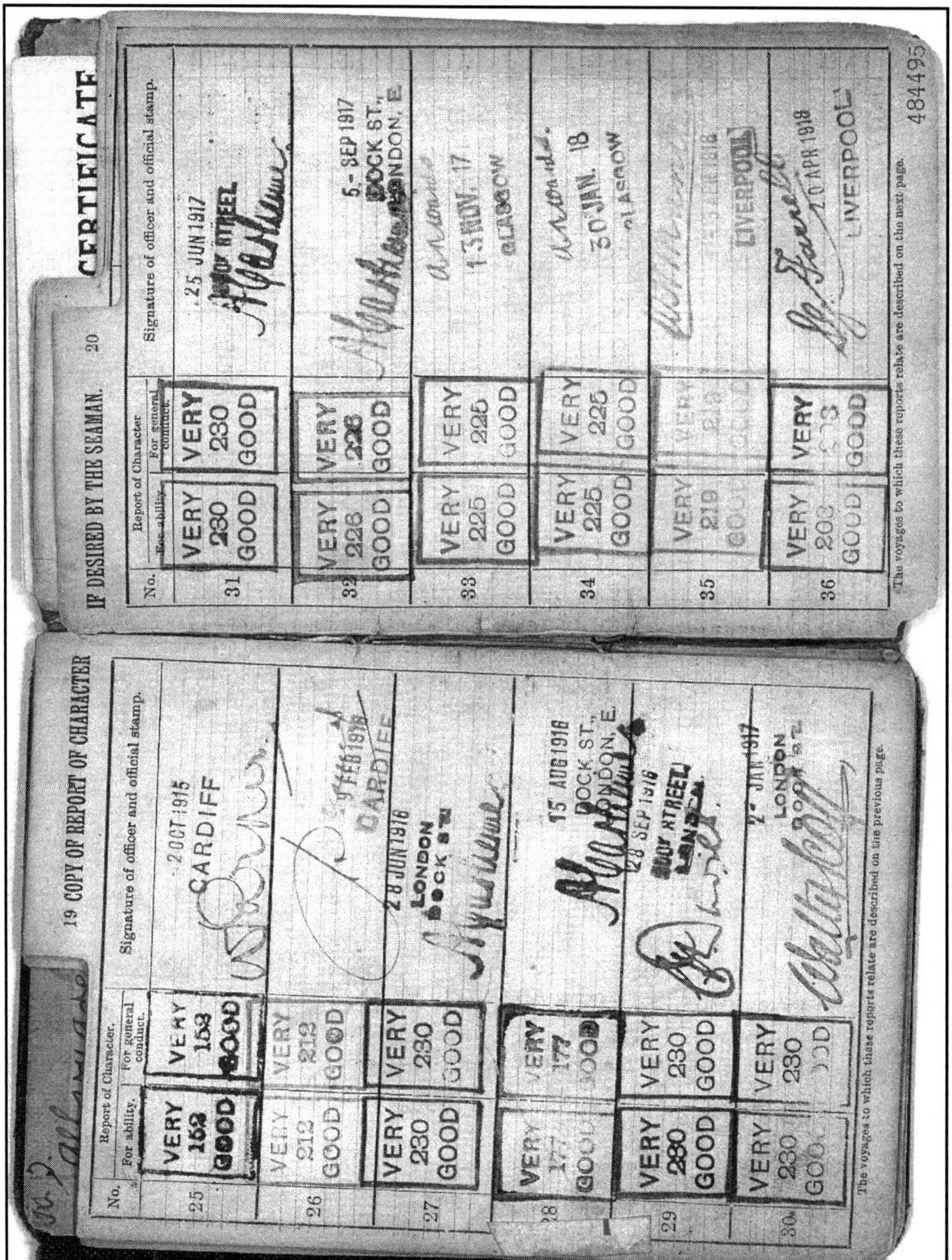

CERTIFICATE

19 COPY OF REPORT OF CHARACTER

No.	Report of Character. For ability.	For general conduct.	Signature of officer and official stamp.
25	VERY 152 GOOD	VERY 152 GOOD	20 OCT 1915 CARDIFF
26	VERY 212 GOOD	VERY 213 GOOD	9 FEB 1916 CARDIFF
27	VERY 230 GOOD	VERY 230 GOOD	28 JUN 1916 LONDON DOCK ST.
28	VERY [illegible] GOOD	VERY 177 GOOD	15 AUG 1916 DOCK ST. LONDON, E
29	VERY 230 GOOD	VERY 230 GOOD	28 SEP 1916 [illegible] STREET LONDON
30	VERY 230 GOOD	VERY 230 GOOD	2 JAN 1917 LONDON DOCK ST.

The voyages to which these reports relate are described on the previous page.

IF DESIRED BY THE SEAMAN. 20

No.	Report of Character. For ability.	For general conduct.	Signature of officer and official stamp.
31	VERY 230 GOOD	VERY 230 GOOD	25 JUN 1917 [illegible] STREET
32	VERY 226 GOOD	VERY 226 GOOD	5 SEP 1917 DOCK ST. LONDON, E
33	VERY 225 GOOD	VERY 225 GOOD	13 NOV 17 GLASGOW
34	VERY 225 GOOD	VERY 225 GOOD	30 JAN 18 GLASGOW
35	VERY 218 GOOD	VERY 218 GOOD	[illegible] 1918 LIVERPOOL
36	VERY 203 GOOD	VERY 203 GOOD	20 APR 1918 LIVERPOOL

The voyages to which these reports relate are described on the next page.

484495

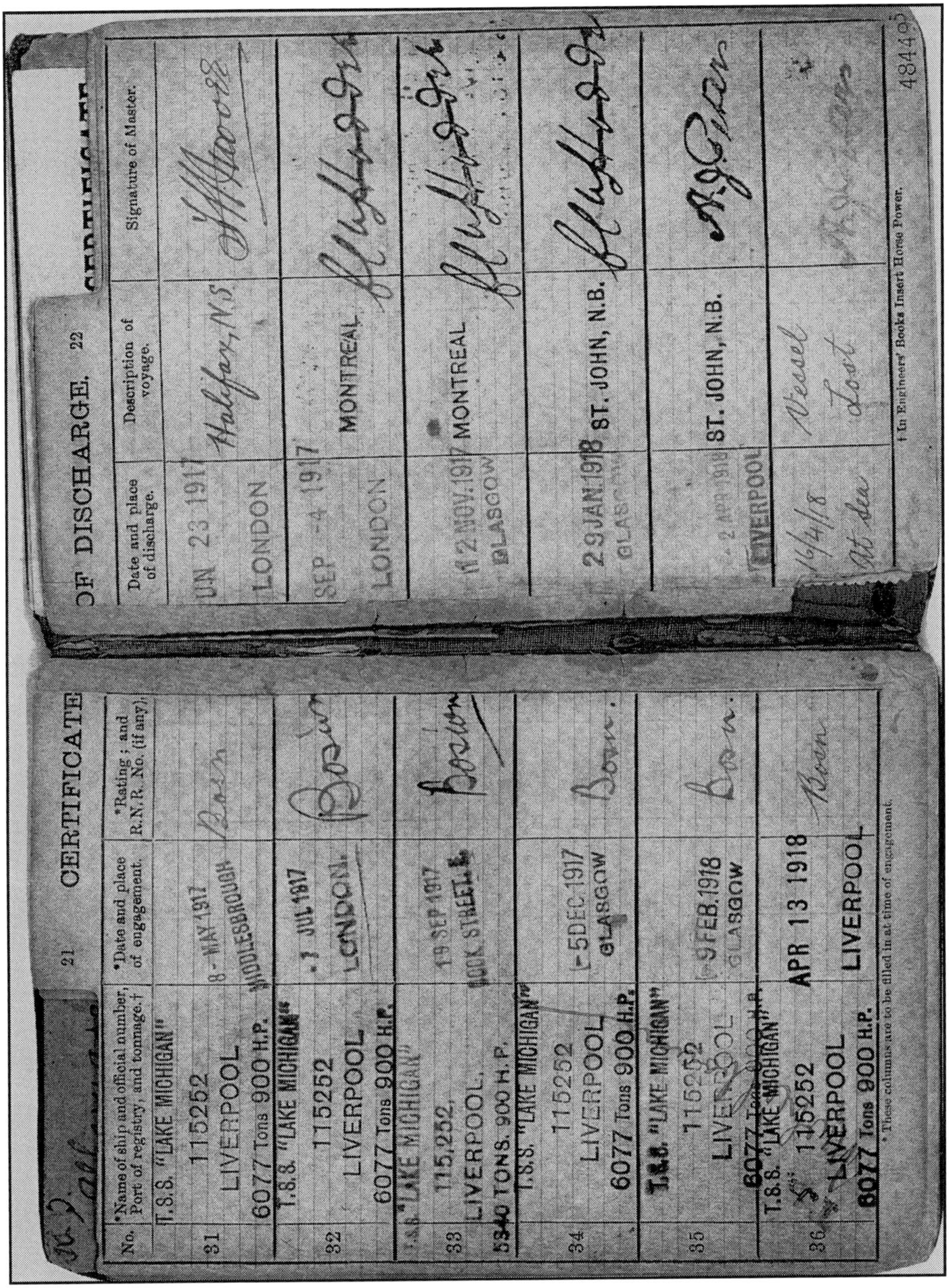

21 CERTIFICATE OF DISCHARGE. 22

No.	*Name of ship and official number, Port of registry, and tonnage.†	*Date and place of engagement.	*Rating; and R.N.R. No. (if any).	Date and place of discharge.	Description of voyage.	Signature of Master.
31	T.S.S. "LAKE MICHIGAN" 115252 LIVERPOOL 6077 Tons 900 H.P.	8-MAY 1917 MIDDLESBROUGH	Bosn	JUN 23 1917 LONDON	Halifax, N.S.	[illegible]
32	T.S.S. "LAKE MICHIGAN" 115252 LIVERPOOL 6077 Tons 900 H.P.	7 JUL 1917 LONDON	Bosn	SEP 4 1917 LONDON	MONTREAL	[illegible]
33	T.S.S. "LAKE MICHIGAN" 115,252 LIVERPOOL. 5340 TONS. 900 H.P.	19 SEP 1917 DOCK STREET, E.	Bosun	12 NOV 1917 GLASGOW	MONTREAL	[illegible]
34	T.S.S. "LAKE MICHIGAN" 115252 LIVERPOOL 6077 Tons 900 H.P.	5 DEC 1917 GLASGOW	Bosn.	29 JAN 1918 GLASGOW	ST. JOHN, N.B.	[illegible]
35	T.S.S. "LAKE MICHIGAN" 115252 LIVERPOOL 6077 Tons 900 H.P.	9 FEB 1918 GLASGOW	Bosn.	2 APR 1918 LIVERPOOL	ST. JOHN, N.B.	R. J. Green
36	T.S.S. "LAKE MICHIGAN" 115252 LIVERPOOL 6077 Tons 900 H.P.	APR 13 1918 LIVERPOOL	Bosn	16/4/18 At sea	Vessel Lost	[illegible]

* These columns are to be filled in at time of engagement.

† In Engineers' Books insert Horse Power.

48445

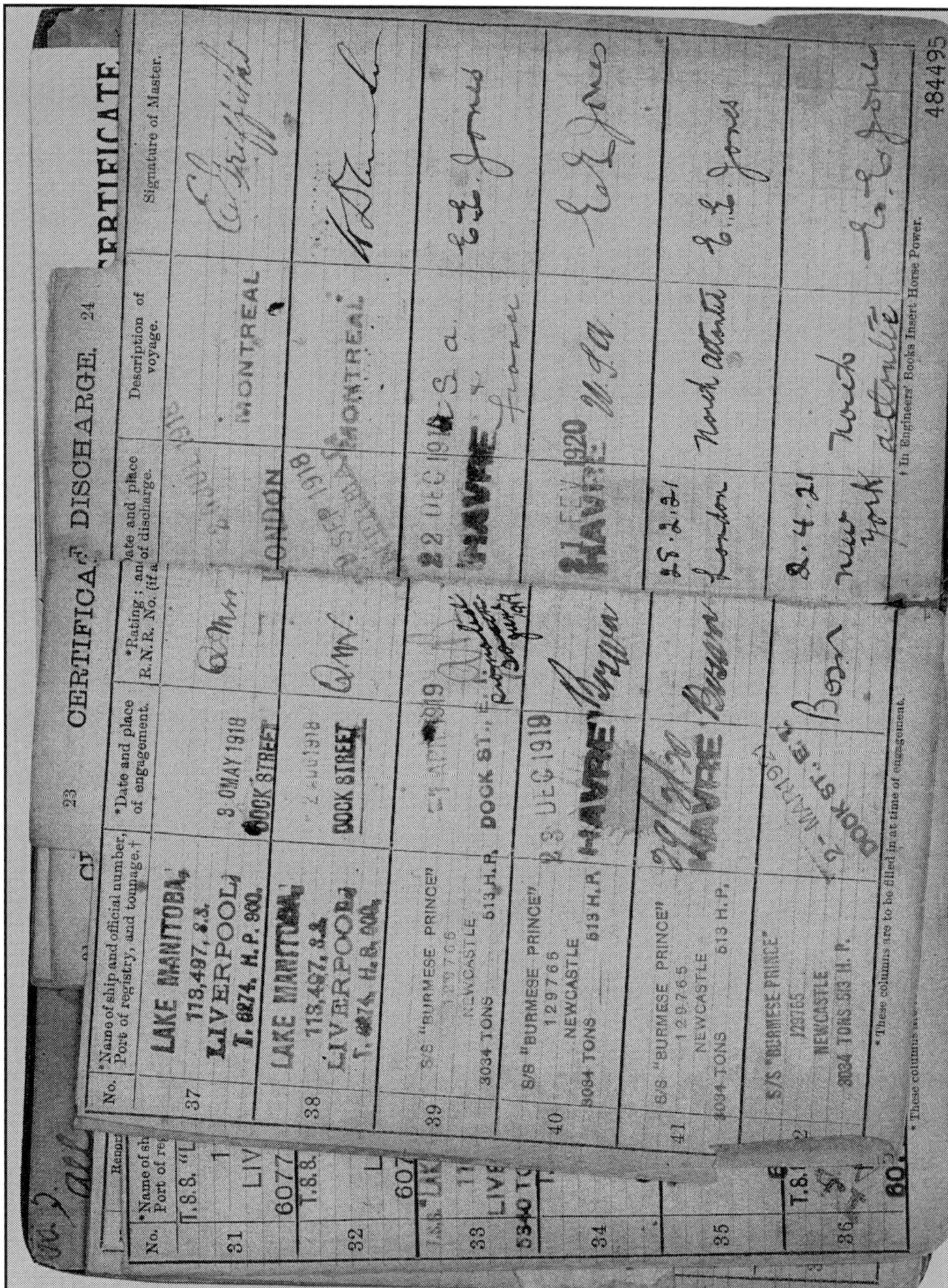

23 CERTIFICATE OF DISCHARGE. 24

No.	*Name of ship and official number, Port of registry, and tonnage.†	*Date and place of engagement.	*Rating; and R.N.R. No. (if any)	Date and place of discharge.	Description of voyage.	Signature of Master.
37	LAKE MANITOBA, 113,497, S.S. LIVERPOOL, T. 6274. H.P. 900.	30 MAY 1918 DOCK STREET	Q.Mr	LONDON	MONTREAL	[illegible]
38	LAKE MANITOBA, 113,497, S.S. LIVERPOOL, T. 6274. H.P. 900.	2 JUL 1918 DOCK STREET	Q.M.	9 SEP 1918	MONTREAL	[illegible]
39	S/S "BURMESE PRINCE" 129765 NEWCASTLE 3034 TONS 513 H.P.	APR 1919 DOCK ST., E.	[illegible]	22 DEC 1919 HAVRE	U.S.A. & France	E. E. Jones
40	S/S "BURMESE PRINCE" 129765 NEWCASTLE 3034 TONS 513 H.P.	23 DEC 1919 HAVRE	Bosun	FEB 1920 HAVRE	U.S.A.	E. E. Jones
41	S/S "BURMESE PRINCE" 129765 NEWCASTLE 3034 TONS 513 H.P.	30/12/20 HAVRE	Bosun	25.2.21 London	North atlantic	E. E. Jones
42	S/S "BURMESE PRINCE" 129765 NEWCASTLE 3034 TONS 513 H.P.	2 MAR 1921 DOCK ST. E.	Bosn	2.4.21 New York	North atlantic	E. E. Jones

* These columns are to be filled in at time of engagement.

† In Engineers' Books Insert Horse Power.

484495

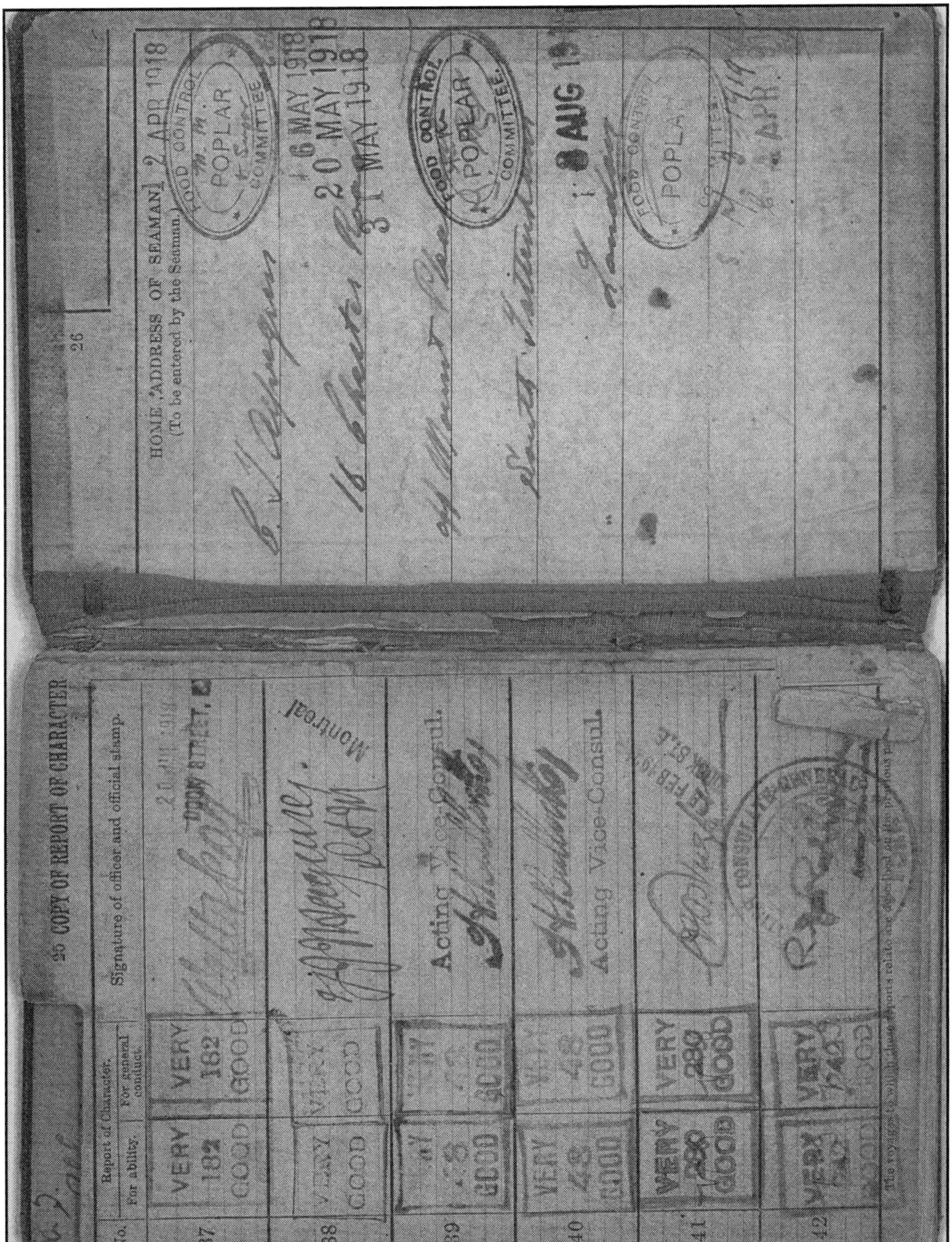
25 COPY OF REPORT OF CHARACTER
No.
Report of Character.
For ability.
For general conduct.
Signature of officer and official stamp.
Acting Vice-Consul.
Acting Vice-Consul.
Montreal
26
HOME ADDRESS OF SEAMAN
(To be entered by the Seaman.)
FOOD CONTROL POPLAR COMMITTEE
2 APR 1918
6 MAY 1918
20 MAY 1918
31 MAY 1918

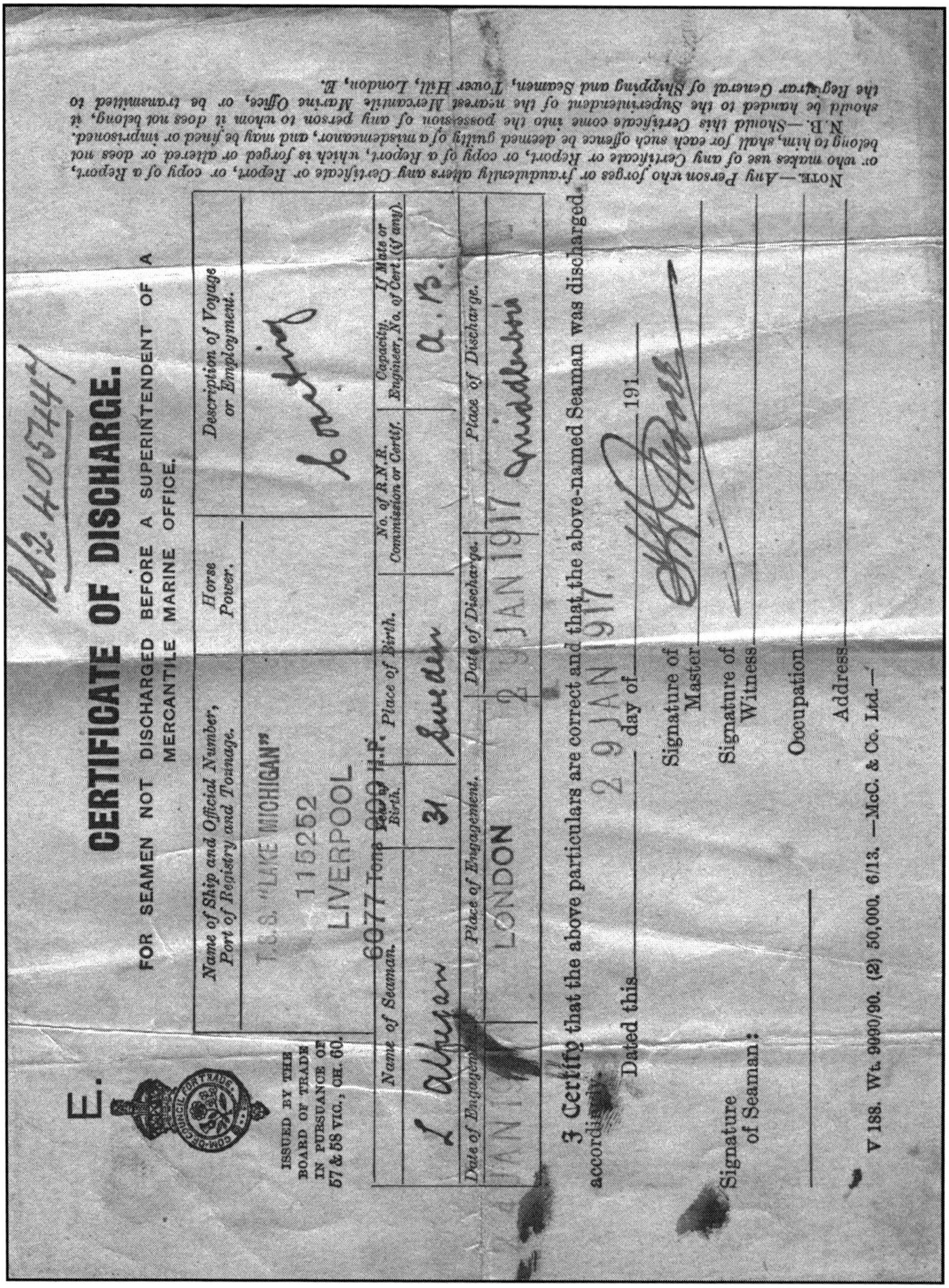

E.

ISSUED BY THE BOARD OF TRADE IN PURSUANCE OF 57 & 58 VIC., CH. 60.

CERTIFICATE OF DISCHARGE.

FOR SEAMEN NOT DISCHARGED BEFORE A SUPERINTENDENT OF A MERCANTILE MARINE OFFICE.

Name of Ship and Official Number, Port of Registry and Tonnage.	Horse Power.	Description of Voyage or Employment.
T.S.S. "LAKE MICHIGAN" 115252 LIVERPOOL 6077 Tons 000 H.P.		Coasting

Name of Seaman.	Year of Birth.	Place of Birth.	No. of R.N.R. Commission or Certif.	Capacity, If Mate or Engineer, No. of Cert. (if any).
[illegible]	31	Sweden		A.B.

Date of Engagement.	Place of Engagement.	Date of Discharge.	Place of Discharge.
JAN [illegible]	LONDON	29 JAN 1917	Middlesbro'

I Certify that the above particulars are correct and that the above-named Seaman was discharged accordingly.

Dated this 29 JAN 1917 day of ______ 191__

Signature of Master ______

Signature of Witness ______

Occupation ______

Address ______

Signature of Seaman: ______

V 188. Wt. 9000/90. (2) 50,000. 6/13. —McC. & Co. Ltd.—

NOTE.—*Any Person who forges or fraudulently alters any Certificate or Report, or copy of a Report, or who makes use of any Certificate or Report, or copy of a Report, which is forged or altered or does not belong to him, shall for each such offence be deemed guilty of a misdemeanour, and may be fined or imprisoned.*

N.B.—*Should this Certificate come into the possession of any person to whom it does not belong, it should be handed to the Superintendent of the nearest Mercantile Marine Office, or be transmitted to the Registrar General of Shipping and Seamen, Tower Hill, London, E.*

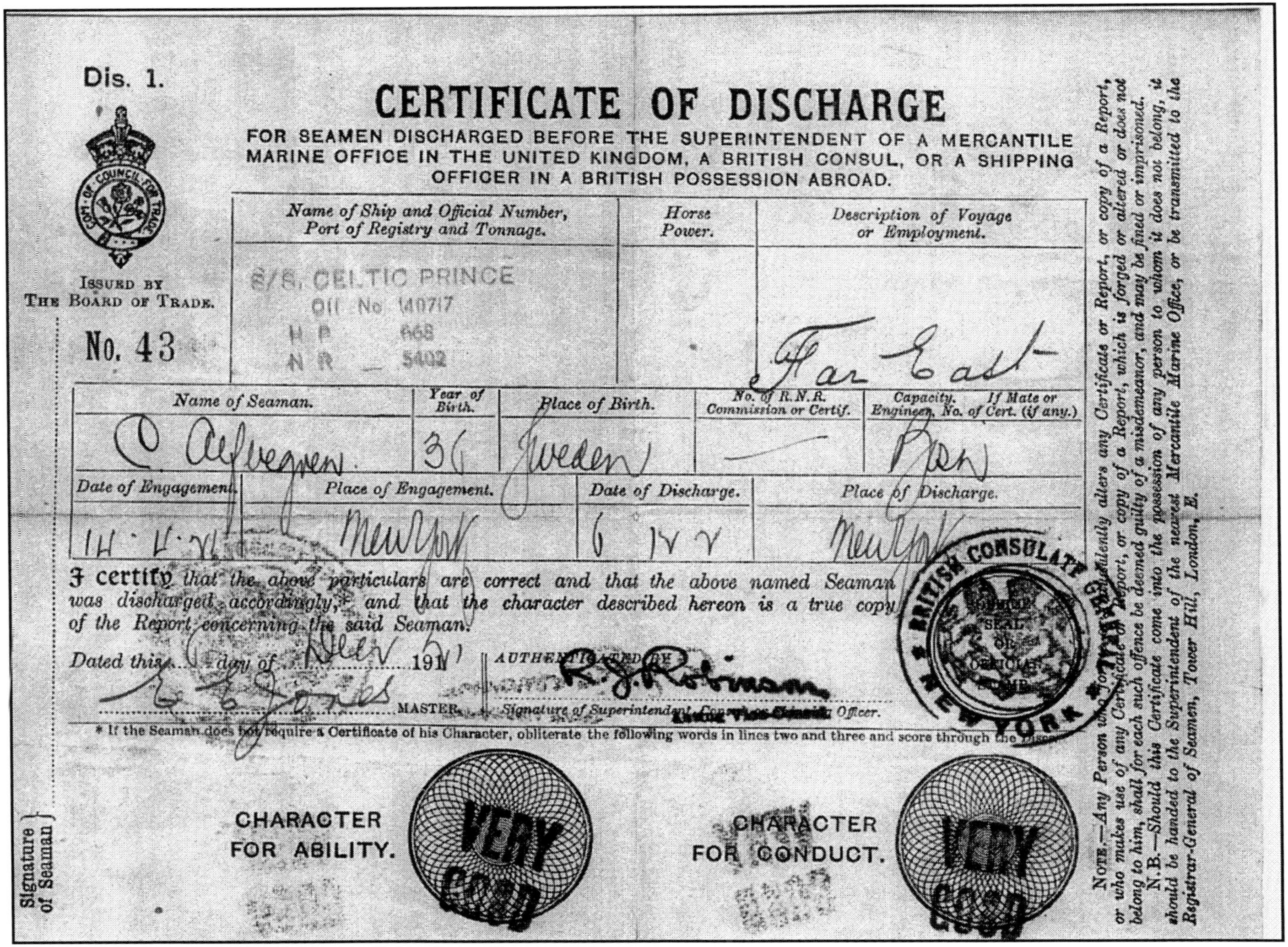

Dis. 1.

ISSUED BY THE BOARD OF TRADE.

No. 43

CERTIFICATE OF DISCHARGE

FOR SEAMEN DISCHARGED BEFORE THE SUPERINTENDENT OF A MERCANTILE MARINE OFFICE IN THE UNITED KINGDOM, A BRITISH CONSUL, OR A SHIPPING OFFICER IN A BRITISH POSSESSION ABROAD.

Name of Ship and Official Number, Port of Registry and Tonnage.	Horse Power.	Description of Voyage or Employment.
S/S. CELTIC PRINCE Off. No. 140717 H P 668 N R — 5402		Far East

Name of Seaman.	Year of Birth.	Place of Birth.	No. of R.N.R. Commission or Certif.	Capacity. If Mate or Engineer, No. of Cert. (if any.)
C Alfvegren	36	Sweden	—	Bos

Date of Engagement.	Place of Engagement.	Date of Discharge.	Place of Discharge.
14 · 4 · 11	New York	6 12 2	New York

I certify that the above particulars are correct and that the above named Seaman was discharged accordingly,* and that the character described hereon is a true copy of the Report concerning the said Seaman.

Dated this 5 day of Dec 1911 AUTHENTICATED BY R. J. Robinson

E. G. Jones ....MASTER. Signature of Superintendent, Consul or Shipping Officer.

* If the Seaman does not require a Certificate of his Character, obliterate the following words in lines two and three and score through the Discs.

CHARACTER FOR ABILITY. VERY GOOD

CHARACTER FOR CONDUCT. VERY GOOD

Signature of Seaman

BRITISH CONSULATE GENERAL * NEW YORK *

NOTE.—Any Person who fraudulently alters any Certificate or Report, or copy of a Report, or who makes use of any Certificate or Report, or copy of a Report, which is forged or altered or does not belong to him, shall for each such offence be deemed guilty of a misdemeanor, and may be fined or imprisoned.

N.B.—Should this Certificate come into the possession of any person to whom it does not belong, it should be handed to the Superintendent of the nearest Mercantile Marine Office, or be transmitted to the Registrar-General of Seamen, Tower Hill, London, E.

Gus's Sailing Routes

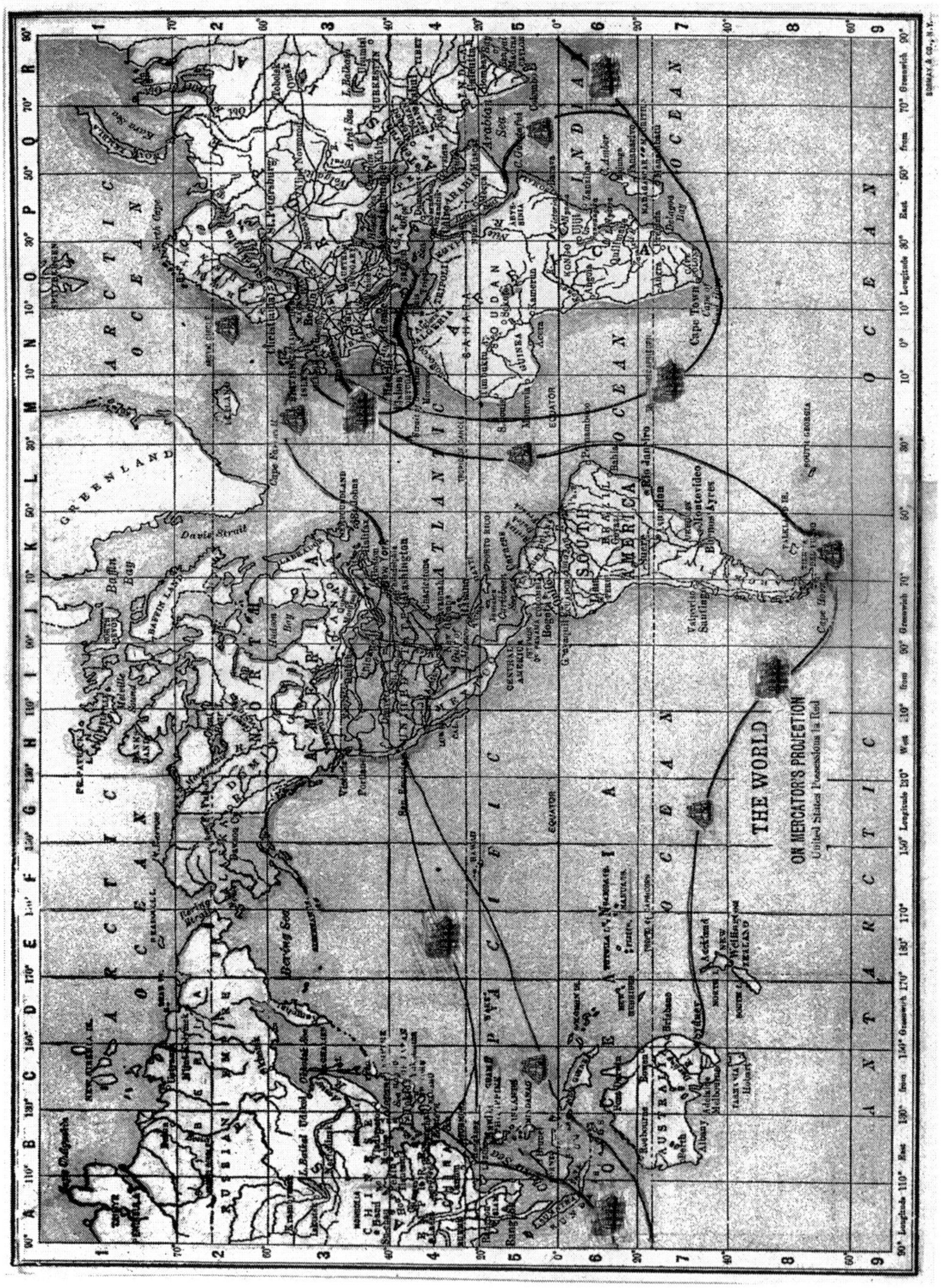
THE WORLD
ON MERCATOR'S PROJECTION
United States Possessions in Red

Nautical Chart

A section of the **Nautical Chart of the Entrance to San Francisco Bay** is on pages 251—254.

The entire chart is on your disc inside the back cover.

To open:

1-Go to "Start" then "My Computer"
2-**Double click** on the CD drive that has the disc in it
3- **Right click** the file CD Nautical Chart

4- a window will pop up> select **Open With**
5- choose **"Windows Picture & Fax Viewer"**

The file should open and you should be able to move around the chart and zoom in or out.

Nautical Chart ***outline of symbols***

from the United States of America Nautical Chart No. 1: abbreviations
NOAA & DOD

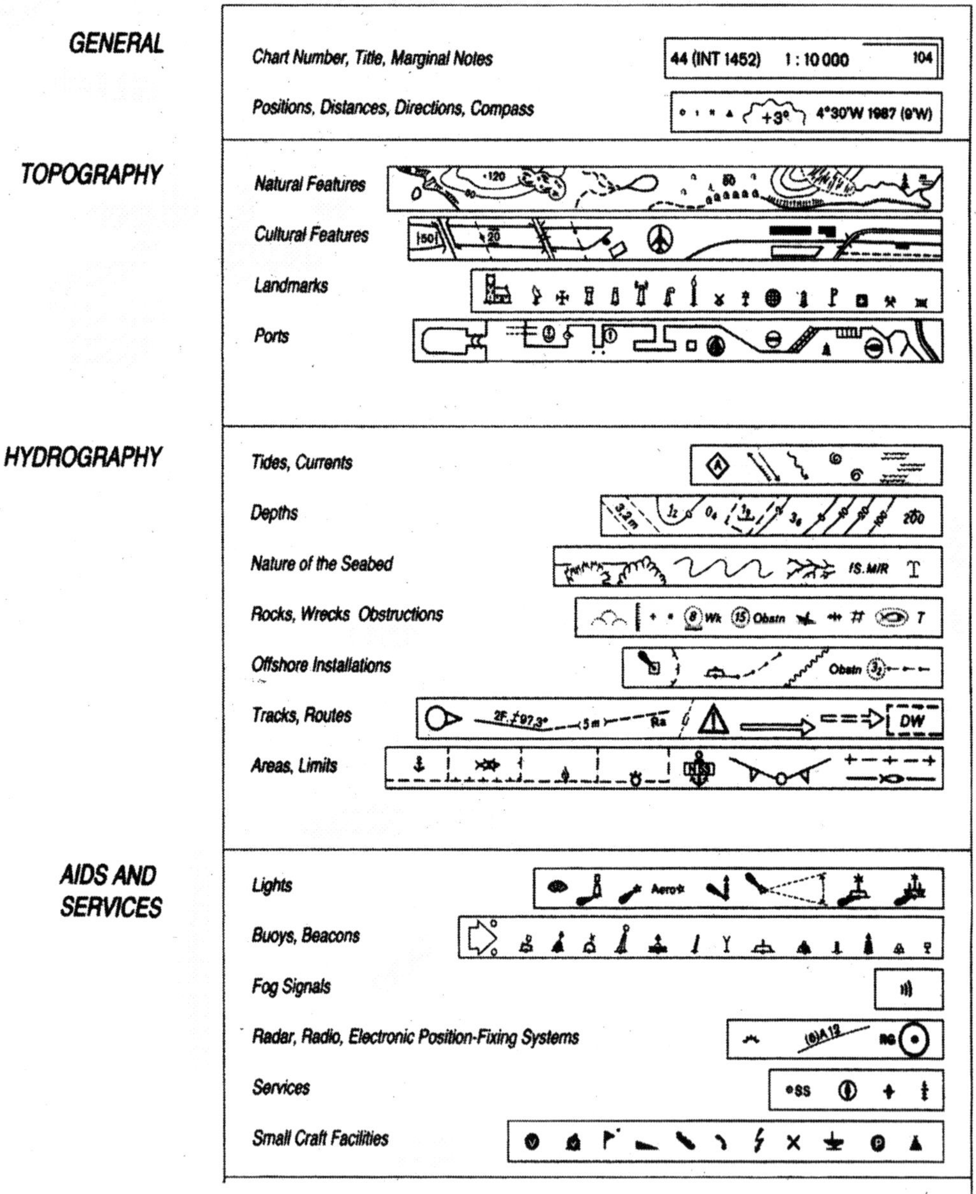

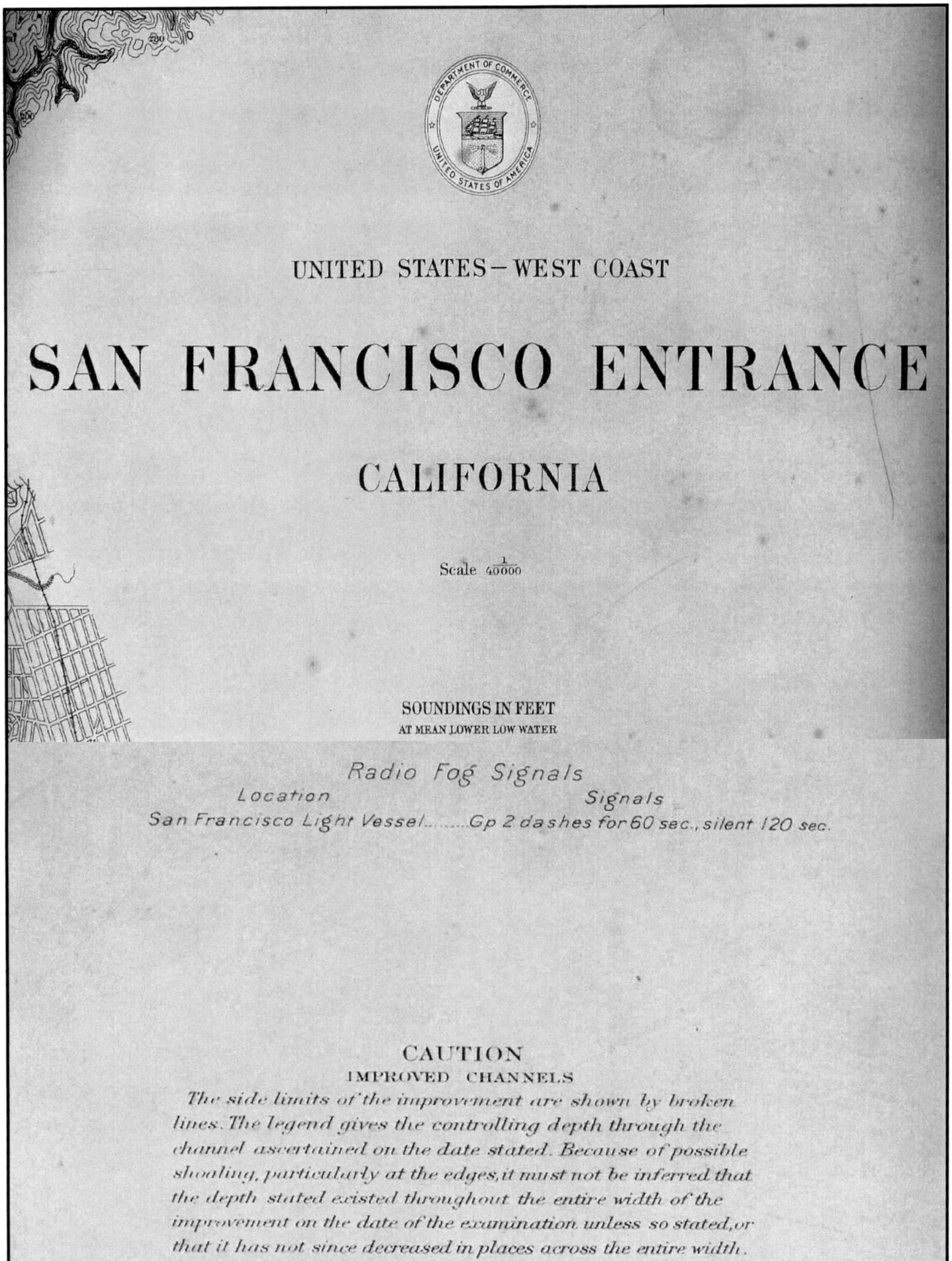
DEPARTMENT OF COMMERCE
UNITED STATES OF AMERICA
UNITED STATES – WEST COAST
SAN FRANCISCO ENTRANCE
CALIFORNIA
Scale 1/40000
SOUNDINGS IN FEET
AT MEAN LOWER LOW WATER
Radio Fog Signals
Location
Signals
San Francisco Light Vessel.........Gp 2 dashes for 60 sec., silent 120 sec.
CAUTION
IMPROVED CHANNELS
The side limits of the improvement are shown by broken lines. The legend gives the controlling depth through the channel ascertained on the date stated. Because of possible shoaling, particularly at the edges, it must not be inferred that the depth stated existed throughout the entire width of the improvement on the date of the examination unless so stated, or that it has not since decreased in places across the entire width.

TIDES

	San Francisco Fort Point	San Francisco Mission St	Oakland	The Brothers Light
High water interval	$11^h\ 39^m$	$12^h\ 07^m$	$12^h\ 15^m$	$0^h\ 15^m$
Higher high water height	5.8 ft.	6.4 ft.	6.6 ft.	6.2 ft.
Lowest tide	-2.5 "	-2.5 "	-2.5 "	-2.5 "

ABBREVIATIONS

Lights: F. *fixed*, Fl. *flashing*, Occ. *occulting*, W. *white*, R. *red*, Alt. *alternating*,
Gp. *group*, Sec. *sector*, m. *miles*, min. *minutes*, sec. *seconds*, ev. *every*, vis. *visible*.
Color white unless otherwise indicated.
Buoys: *C. can N. nun S. spar W. white*
C.G. *Coast Guard Station* N.R.C. *Naval radio compass* R.F.S. *Radio Fog Signal*
M. mud, S. sand, G. gravel, Sh. shells, P. pebbles, Sp. specks, Rk. Rock
bk. black, yl. yellow, gy. gray, bu. blue, dk. dark, lt. light
hrd. hard, sft. soft, fne. fine, crs. coarse, rky. rocky, stk. sticky, brk. broken

AUTHORITIES

Surveys to 1926
Surveys by U.S. Engineers to 1926 and other sources

HEIGHTS *in feet above high water*

RULES OF THE ROAD for Inland Waters
must be followed for harbors inshore of a line through Mile Rocks and Point Bonita Lighthouses

STORM WARNINGS
are displayed by the U.S. Weather Bureau at Point Lobos and San Francisco

24'
22'
122°20'
46'
44'
37°42'
China Basin
Mission Rock
GENERAL
FORBIDDEN ANCHORAGE FOR EXPLOSIVES
Potrero
Potrero Pt
GAS TANK
Islais Cr.
Cable area
BOOM
Butchertown
STACK
BOOMS
Pile
Dry Dock
Hunters Point
Pt. Avisadero
Piles
Bayview
Double Rock
Shag Rock
Candlestick Pt.
Bayshore
TANK
TANKS
Visitation Pt.

18'
16'
Cable and pipe
Alaska Basin
A L A M E D A
TIDAL CANAL
San Leandro Bay
RED TANK
Bayfarm Island
B A Y
46'
44'
37°42'

Nautical Chart